Anxiety and the Equation

Anxiety and the Equation

Understanding Boltzmann's Entropy

Eric Johnson

The MIT Press
Cambridge, Massachusetts
London, England

This book was set in Stone Serif by Westchester Publishing Services.

Library of Congress Cataloging-in-Publication Data is available.

ISBN: 978-0-262-03861-4 (hardcover), 978-0-262-54661-4 (pb)

For those of us who have a tendency to think like Boltzmann (and for lovers of parentheses)

Contents

Acknowledgments

I thank my talented wife, Tricia Johnson, for her help with many of the figures in the book; Nik Money, Diana Davis, Gene Kritsky, Vinay Ambegaokar, and Mark Rance for generously sharing their advice and expertise with me; and Stephen G. Brush for providing me with an unpublished translation of Boltzmann's 1877 paper. I also thank David Lindley, whose book *Boltzmann's Atom* was a valuable source of biographical information, as well as Jermey Matthews, Virginia Crossman, and the other members of the MIT Press, who were enormously helpful throughout the review and production of this book.

Introduction

This book is not a textbook, but there's a chance that you might learn something from it. It's not really a biography either, although it has a lot to say about a man named Ludwig Boltzmann. It's not even a work of nonfiction in its entirety. It follows the historical record, but also knowingly enters places where that record is incomplete and borrows occasionally from the tools of fiction. Only once will this book lie to you, and only then at the end, by which time you will have already learned the truth. The truth in this case is an unpleasant thing, so let's dispense with it first.

1 Boltzmann Kills Himself

It was an inelegant death. Hanging there. He was a fat man. He might have made a convincing Santa Claus on a good day. But the past several years had brought him very few good days. Imagine a Santa Claus who, at the end of a long and successful career, finds himself unable to face the impending arrival of yet another December 25. So too must Ludwig Boltzmann have felt at the age of sixty-two in the summer of 1906.

It was late afternoon, approaching early evening, on September 5 of that year. He had traveled to the coastal village of Duino with his wife and three daughters. Family, friends, and colleagues all knew that he was ill. His eyes no longer worked. He had asthma. He suffered chest pains and headaches and enjoyed no peace of mind. He was a wreck. He had moved from one university to the next with occasional stops at a sanatorium and was diagnosed with neurasthenia, a disorder long since forgotten but one that included symptoms of anxiety and depression. Officials at the University of Vienna thought it unlikely that he would be able to continue his teaching duties, so a younger professor was told that he should prepare to teach physics the following year. Yet everyone still hoped for a recovery. The man inspired little confidence, but

Boltzmann was his country's most accomplished scientist and a beloved teacher. Everyone hoped that a few weeks by the sea would calm his nerves and restore his health, allowing him to return to Vienna for the start of the academic year.

But Boltzmann had already lived more days than he could bear. And though he had planned to return to Vienna the following day, he decided that he would rather kill himself instead. So as his family went down to the sea to bathe, Boltzmann set about the task of committing suicide. He fastened a noose to the crossbar of a window and checked that it would hold him (no small task for that crossbar). Then he placed the noose around his neck, stepped off the ledge, hung there for a few seconds, and died.

He was a kind man with a generous mind. This is his story and the story of his second law.

2 Boltzmann Is Buried (Not Once, But Twice)

Boltzmann's body was returned to Vienna. His obituaries recognized what was widely known: that he was a brilliant but troubled man. Even his chief adversary, Ernst Mach, turned contemplative and conciliatory in his tribute, wondering if a life in science demanded too much of its practitioners, if excessive competition was now to be the norm (an ill-timed sentiment given the fact that the two men's professional grievances had always been personal).[1] If Boltzmann saw little sympathy in life, he received an abundance of it in those obituaries. But he was then quickly forgotten. The goodwill, in fact, seems to have lasted no more than a few days. Only two physicists, Gustav Jäger and Stefan Meyer, attended his funeral the following week.[2]

Boltzmann had apparently faded from prominence, and the paradigm was about to shift. Recent papers by Max Planck and Albert Einstein had provided Boltzmann with all the proof that he needed in his long-standing battle with Mach, but he was too discouraged and disengaged to have even noticed. If history had been kinder, he would have found in those papers his justification. He would have seen the beginnings of a new quantum century and a new modern physics. He would have seen himself a modern man. But he killed himself instead. And so he was put in a grave, and that grave fell into disrepair, until twenty-three years later when the good people of Vienna decided that their native son deserved better.

His coffin was removed from its original resting place, albeit with some difficulty. The burial plot was no longer his alone, but something that he shared with another occupant.[3] It therefore was not enough to dig down in order to retrieve Boltzmann. It was necessary to dig at an angle so as not to disturb the other body. (Boltzmann was never one to make things easy, not even in death.) His coffin was successfully removed, however, and taken to a place of honor in Vienna's Central Cemetery, where he was buried alongside Beethoven, Brahms, Schubert, and Strauss. Always a great lover of music, Boltzmann was now among his heroes. And the monument that he was given could not have portrayed a prouder image of the man. It included a bust of white marble, the outlines of which elevated him to a physical ideal never seen in his life. His jaw was sharp, his hair and beard neatly groomed. It was not a faithful rendering. And why did he appear so stern? Boltzmann was many things in life: warm, especially among family and friends, and sometimes awkward or prickly, but never stern, although maybe there's no harm in remembering him this way. Perhaps we simply need our heroes to look like heroes. But then we miss an opportunity to know the real Boltzmann and never realize that no great distance separates us from this man.

Boltzmann was a man of reason who suffered, often for no good reason and of his own doing. He was a man with serious adversaries, but none so formidable as the habits of his own mind. Yet he managed to create great things. And chief among his many accomplishments is the equation that is inscribed atop his monument:

$S = k \log W$

If we are to understand Boltzmann, we need to understand this equation.

3   Start Simple

The equation is simple. Or is it? It says that the entropy S is equal to a constant k multiplied by the logarithm of the number of permutations W. Entropy? Permutations? These are not simple things. The equation is simple only in the sense that it can be written compactly. It seems that the equation defines a thing called entropy, and entropy is widely regarded as something strange and mysterious and misunderstood—decidedly not simple. It is also regarded, at least by those of us who are inclined to think about such things, as important—important because it points us in the right direction, namely, the future. Boltzmann did not give us a crystal ball. He gave us something much more valuable: a modern view of the second law of thermodynamics, one that recognized the existence of atoms and the utility of statistics, ideas that sparked visceral opposition in the nineteenth century but proved necessary and commonplace in the twentieth. The second law tells us that the entropy of an isolated system remains constant or increases with time. It says the thing that we call S tends toward a maximum value. If this maximum value has already been reached, S remains constant. And if not a maximum, its value will surely rise to a maximum value later. The second law acts as an arrow of time. Boltzmann was not the first person to propose this idea, but he was the first to provide a credible reason why.

If we are to understand that reason, we first need to define the system of interest. The second law refers to an *isolated* system. By stipulating that the system is isolated, we know how it interacts or, in this case, does not interact with its surroundings. An isolated system is isolated in the sense that it does not exchange matter or energy with its surroundings. No particles are allowed in or out, and the boundary is fixed and allows no heat to cross. No system is truly isolated, but that's the scenario to which the second law applies. Imagine, for example, that our system is confined to a really good thermos bottle.

We then need to identify the contents of the thermos bottle. Boltzmann would have us place a gas of some sort inside it. Gases were his preferred subject, his most famous work being a book called *Lectures on Gas Theory*. But how should we think about a gas? How should we picture it? If we are to trust our eyes, we should picture a mostly empty space. The air around us, the gas with which we're most familiar, certainly looks empty. But our eyes can be deceived because scattered about this mostly empty space are particles that are too far apart from one another to amount to anything visible. We can simplify things even further by insisting that the particles are smooth, featureless spheres, lacking any internal structure or movable parts. And although these particles can move about and collide with one another, they don't otherwise stick to or repel one another. Imagine a collection of tiny (very tiny) billiard balls.

It's a simple system. It might, in fact, appear too simple. Modern physics admittedly offers more exotic objects for our consideration: strings in high-dimensional spaces, black holes lurking enormous distances away. A gas probably seems rather mundane in comparison. And we might be led to believe that Boltzmann's *Lectures on Gas Theory* is a relic or (at best) something from a quaint, forgotten past. But there's good reason to focus on something simple like a gas. It's by studying something simple that we can begin to understand this thing called entropy that's supposedly very complicated. We need to start somewhere. And sometimes it makes sense to start at the beginning.

4 Before Things Got Weird

I selected a relatively simple thing, a gas, as our system in the hope that it will be relatively simple to understand. We'll take the view here of an omniscient observer (a convenient perch on which to sit). If we had access to all that we might want to know about the gas, what information would we find to be most useful? A picture would certainly help: it would tell us where to find the particles and give us their positions. But a movie would be even better. It would tell us not only where to find the particles, but also where they're going and how fast they're moving. It would give us their velocities.

Here we are confronted by the cries of those familiar with Herr Heisenberg and his uncertainty principle. They tell us (quite correctly) that the information provided by this movie is a dream that contradicts the reality of quantum mechanics. No movie can provide us with the positions and velocities of the particles, at least not with any certainty. This sad fact is not some technological shortcoming; it's a fundamental limit of what we can know. At smaller and smaller scales, things behave less and less like hard, distinct objects. Things get fuzzy; positions and velocities become ill defined. So says the uncertainty principle. But we cannot fault Boltzmann for not having anticipated this principle. He was much too busy building the early foundations of quantum mechanics to have foreseen all of its many consequences. He worked at a time when common sense was still a reasonable guide and when particles and other objects, however small, were seemingly real things, consistent with our everyday experience.

Today such a view is called classical to distinguish it from the quantum model that replaced it. And in what follows, we'll adopt a classical view of the world. We'll allow ourselves the positions and velocities of the particles. And we should not feel too guilty. A gas is very well approximated by classical physics, and a classical gas is the natural starting point in many undergraduate textbooks. (So we're in good company.) Our goal here is not to account for all of physics. It is to understand entropy in the context of a few common scenarios.

5 Postmortem Psychiatry

Everyone loves a troubled genius. It's the stuff of popular legend. Van Gogh was not just brilliant. He was crazy and brilliant. He cut off his own ear. Our hero Boltzmann managed to hold on to both ears, but his story is no less dramatic and the end of his life no less tragic. Here we will explore his struggle in the hope that we might understand the troubled mind that managed to make sense of the second law. Of course, it doesn't really matter that he struggled. Entropy would be no less significant an idea if it were the product of a stable psyche. But for some strange reason, we seem to wish mental illness on our geniuses. Why? We might simply be jealous. We know that Boltzmann was blessed, his intellect rising far above the masses. Some dark shadow part of our brains might resent what he was given. He inherited an extreme position on the bell curve, and for that, we might deep down want him to have suffered. We might want blessings and curses to be distributed in equal measure. But jealousy need not be tainted with any malice. Boltzmann's suffering might actually help to keep our jealousy in check by offering us certain reassurances. We might not be exceptional, we tell ourselves, but at least we're well adjusted and fully functional. Maybe we should wish for nothing more than an unremarkable life. Or maybe it has nothing at all to do with us personally. Perhaps we simply recognize that creative and destructive tendencies are common companions and often find ample resources in the same person.

But, still, we love a troubled genius. And although psychiatric textbooks provide a long list of possible diagnoses, we seem to favor disorders whose symptoms are the most brazen. Schizophrenia and bipolar disorder readily capture our imaginations. These disorders are deserving of a great mind, and they inspire screenplays. Boltzmann's example serves as fodder for this notion. Conventional wisdom asserts that he suffered from bipolar disorder, a claim that's often accompanied by a remark that he made on the occasion of his sixtieth birthday. He joked that his personal struggles could be traced to the circumstances of his birth. He arrived into this world as Mardi Gras turned to Ash Wednesday, in a house in which there was also a dance hall. And so his first cries were made "amid the dying noises of the dance."[1] Little baby Boltzmann apparently witnessed in those early hours a mood swing that was imposed by the liturgical calendar, but he would eventually learn that internal forces could also turn against him. Fortunately, however, he was spared that realization until much later in life.

He was a happy and successful young man: a doctoral degree at the age of twenty-two (just three years after enrolling as an undergraduate), a position as chair in mathematical physics at the age of twenty-five (one of a very small number of such jobs available at the time), a doting wife (who conveniently replaced his doting mother), and a home that was filled with all of the things that he loved (food and drink, a piano, four children whom he loved and who adored their papa).

It was not until middle age that Boltzmann first faltered. When he was forty-four years old, he was offered a highly prestigious position at the University of Berlin. A move from Graz, where he was currently employed, to Berlin represented a major advance in his career. In Graz, the quality of his science had been remarkably high, but the stakes had also been relatively low. Boltzmann had often complained that he felt isolated there, but isolation somewhat suited him and may even have facilitated his early success. If he were to accept his rightful place among the titans of Berlin, he would need to defend his work in a much more highly charged arena. Berlin would present him with personal challenges from which he had previously been spared.

In some parallel world, one not traced by our own historical record, Boltzmann might have welcomed this opportunity. He was naturally ambitious and fully aware of his talents. But he was not built for Berlin, which was clear when he visited the German capital to negotiate the terms of his contract. During that visit, Boltzmann dined with Professor Hermann von Helmholtz and his wife, Anna. Boltzmann was likely giddy at the prospect that he would soon count Helmholtz, the *Reichschancellor* of German physics, as a colleague. But during the meal, he committed that most egregious of social missteps: he picked up the wrong piece of cutlery (god forbid). His hostess duly noted his gaffe and said to him: "Herr Boltzmann, in Berlin you will not fit in."[2] Now if she had been a better hostess, Frau Helmholtz would have ensured that her guest did in fact fit in. Instead she threw a gibe, which might have also been a kind of warning. But she was right, regardless of her intentions. Had Boltzmann moved to Berlin, he would not have fit in. Berlin bound he was not. And in the course of not moving to Berlin, Boltzmann began to unravel.

His first serious mistake, outside the faux pas at the Helm-holtz dinner table, was to sign a letter of intent in Berlin without first notifying the Austrian authorities that he would be leaving Graz. As rumors inevitably began to circulate, he was informed that the Austrian education minister "would regret it most keenly if your honor's excellent services were to be lost to the University of Graz and to the Fatherland, and would place great value on your honor's remaining in his present position."[3] Boltzmann provided verbal reassurance to a ministry official that he had not yet made a decision, but it was suggested that he make that position clear in writing. At that point, he must have recognized the severity of his predicament. Boltzmann had managed to mislead both the Prussian kaiser and the Austrian emperor at a time when academic appointments were matters of national pride. Professors did not consider *offers* for open positions; they responded to *calls* from a royal court.[4] Boltzmann was not equipped to navigate these waters.

He first asked that the offer be revoked on account of his poor eyesight, but he would not get off so easily.[5] The Prussian ministry responded that his myopia was something that they could certainly accommodate. Meanwhile, the Austrian ministry offered to increase Boltzmann's salary in order to keep him in Graz. He was, it seems, the subject of a bidding war. He wrote again to Berlin, this time to suggest that he was seeking release from the Austrian ministry (which, so far as we know, he was not), but that letter was quickly followed by two others of June 24, 1888, where he changed course again.[6] He claimed in those letters to lack the mathematical expertise that was needed for the job, which was a sad and ridiculous proposition and one that indicated an increasing sense of desperation. (Who, if not Boltzmann, would have been qualified to serve as a professor of mathematical physics?) He also included with these letters reports from both his eye doctor and his psychiatrist. The eye doctor confirmed that Boltzmann's vision was poor and worsening, and the psychiatrist reported that he suffered from neurasthenia and general nervousness.[7] It seems that Boltzmann's solution to his problem was to deliberately portray himself as intellectually, physically, and psychologically unfit.

Things were about to get worse. Three days later, in a telegram to Berlin, he asked that the letters of June 24 be returned to him unopened. But he floundered yet again the following day and suggested in another telegram that the letters should be opened after all.[8] It was getting ugly. Prussian officials now had reason to believe that Professor Boltzmann lacked more than the necessary table manners to succeed in Berlin, so they contacted his wife, Henriette, through a family friend and asked if Boltzmann had in fact sought an official release from his position in Graz. Henriette wrote back, admitting that he had not. She knew her husband well and clearly saw her unwelcome future in that letter, writing: "I fear, unfortunately, that he will feel unhappy for the rest of his life to have forfeited forever a position that would so well suit his preferences. There is hardly anything that can be done!"[9]

The Prussian kaiser, however, knew what needed to be done and released Boltzmann of his obligation on July 9, allowing the man to return to his happy and successful routine in Graz. But it was not to be. Boltzmann immediately responded to the Prussian ministry, describing in embarrassing detail how he had struggled with their negotiations. He asked that they send back a telegram with the words "too late" or "still time."[10] It was a cringeworthy spectacle, and it was not yet over. The Prussian ministry confirmed that it was in fact too late, but Boltzmann wrote again to Berlin the following fall, indicating that the Austrian ministry had given him a pay raise. He apparently thought that the negotiations might still be open, which they decidedly were not, and hoped to convince the Prussians that he was still desirable. He asked if the "forfeited" position might still be available.[11] The answer of course was negative. The Prussian ministry kindly (and wisely) recommended that Boltzmann retain his current position.

Did Boltzmann accept defeat (which he had clearly earned)? Had he endured enough humiliation at his own hands? Unfortunately not. His last desperate outreach was to Helmholtz himself. In December of that dreadful year, he sent Helmholtz a paper that he had written. The paper, he admitted, was not an especially significant contribution, but he hoped that it would demonstrate that his "overstrained nerves" had subsided.[12] He expressed his regret at the lost opportunity and asked if Helmholtz might help him in any way. But it's not known how (or if) Helmholtz may have replied.

The arc of a great life was now on a downward trajectory. Boltzmann had managed to avoid failure for the first forty-four years of his life, but when he discovered failure, he failed fantastically. His performance in this debacle, perhaps the most badly bungled job negotiation in the history of the academy, forever changed his life. Failure introduced him to self-doubt, which was now to be his constant companion.

6 But How Crazy?

It sometimes helps to ask an admittedly crazy question. Let's say that you walk into an auditorium where a large crowd has assembled and note that everyone is sitting on the left-hand side of the room. You look for a familiar face, hoping to ask someone for an explanation, but you see a room full of strangers. You suspect, however, that you know why the crowd displays such a strong preference for one side: you suspect that all of the air has migrated to the left side of the room, leaving the right side uninhabitable. You consider approaching someone to ask if he or she might confirm your suspicions, but you hesitate, common sense getting the better of you. Maybe your explanation is not such a reasonable one. What are the chances that all of the air particles would isolate themselves to one side? It's a crazy question. But how crazy? Is it absolutely ridiculous? Or just a bit odd?

It would help if we could put a number, a measure, on crazy. The scenario that I described is certainly a low-probability event, but we need to decide if it's absolutely unthinkable or just highly unlikely. We need a number. Without one, we can say only that the scenario is not a cause for legitimate concern (because people of sound mind have found no reason to worry about such things). Our world is not one in which the air spontaneously conspires against us. If you decide to move from one side of the room, where the air is known to be plentiful, to the other side of the room, you can confidently assume that the air will be just as plentiful on the other side. Predictions of this sort are highly reliable; we would find it difficult to function otherwise. But let's find the number anyway. Let's put a number on crazy. Why? Because useful results (things like the second law) are sometimes found in (seemingly) not so useful places.

We'll follow our guiding principle: start simple. Air is actually a mixture of several different elements and compounds, most notably nitrogen, oxygen, argon, and carbon dioxide. And the number of air particles in a room of any size is staggeringly large, a fact that will ultimately prove to be very important but might initially stand in our way. So rather than tackle a room full of air, let's work with a smaller number of simpler particles. Let's consider a room with just two identical atoms, both of which are smooth and featureless. And let's assume that these atoms don't stick to or repel one another and therefore satisfy all of our simplifying assumptions.

If these two atoms are the relevant players, then we need to outline their field of play. In order to keep things simple, let's divide the room up into just two sides, labeled left and right. Although no actual physical barrier separates the two sides, the labels left and right provide a rough description of the atoms' positions. (We'll later consider the atoms' velocities.) For now, we want to know the likelihood, the probability, that both of the atoms are on the left-hand side of the room. We can answer this question by counting up all of the ways that the atoms can be arranged. In doing so, we find that the scenario just described, with both atoms on the left, is one of four possible configurations (figure 6.1):

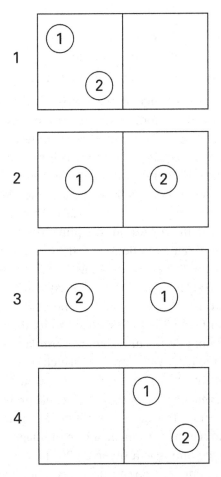

Figure 6.1
Four possible configurations for
two atoms

1. Atom 1 on the left, atom 2 on the left
2. Atom 1 on the left, atom 2 on the right
3. Atom 1 on the right, atom 2 on the left
4. Atom 1 on the right, atom 2 on the right

Which of these configurations is most likely? Or is there a most likely configuration? In the absence of any additional information, it seems reasonable to assume that the four configurations are equally likely, in which case there is a one in four chance of finding both atoms on the left-hand side. The assumption of equal probabilities is an old idea, one that preceded Boltzmann. It has gone by many different names: the principle of insufficient reason, the principle of indifference, the principle of fair apportionment, the equal a priori probability postulate. These names all suggest that we should not, unless given good reason, play favorites. All configurations are created equal, and no configuration is more or less likely than any other configuration. No less an authority than Wikipedia calls this assumption the fundamental postulate of statistical mechanics, the field whose origins I sketch here.

There remains, however, a key difference among the configurations even when we assign them equal probabilities. In order to recognize this difference, we need to grant ourselves the ability to see the very small, a leap of imagination that's allowed us, the omniscient observer. But here we encounter limits even to our keen powers of observation. (Omniscience was apparently fleeting.) Consider the following test. If we were to turn our backs on the system, allow the system to randomly sample one of the four configurations available to it, and then turn back around, could we correctly identify which of the four configurations had been sampled? We would certainly know if both atoms were on the left (configuration 1) or if both

atoms were on the right (configuration 4). Those two configurations are easily recognized. The other two configurations,
however, are problematic. We need to ask ourselves what we
would see if either of these two configurations (configuration
2 or 3) had been sampled. We would see one atom on the left
and another on the right. But would we know which of the
two atoms, 1 or 2, was on the left and which on the right?
Would we be able to distinguish between configurations 2
and 3? The answer depends critically on whether we can distinguish between atoms 1 and 2. I introduced the labels 1
and 2 for accounting purposes. They were needed in order to
enumerate all of the possible configurations. But we should
not hold on to these labels any longer than necessary, for the
simple reason that microscopic objects don't come with little
microscopic labels attached to them. The labels reflect no
inherent difference between the two atoms. And without real
labels, without some identifying feature, the two atoms are
indistinguishable. And indistinguishable atoms imply indistinguishable configurations. Configurations 2 and 3 look the
same and therefore lead to one recognizable outcome. By dis-
carding the labels, we find that configurations 2 and 3 are
lumped together.

It might help at this point to introduce some formal language. Each of the four possible configurations is called a
microstate. Some microstates, such as those represented by
configurations 1 and 4, can stand on their own. Removing the
labels from either of these microstates gives a unique, recognizable outcome. Other microstates, however, are lumped together
because they look the same after the labels have been removed.
This process of removing the labels and lumping together
convergent microstates leads to what are called macrostates.
Each recognizable outcome of our thought experiment then

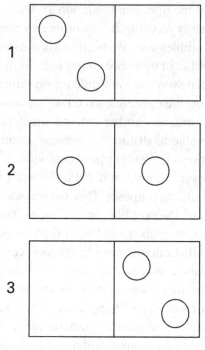

Figure 6.2
Three possible macrostates for
two atoms

corresponds to a macrostate, of which the current system has
three (figure 6.2):

1. Two atoms on the left, zero atoms on the right

2. One atom on the left, one atom on the right

3. Zero atoms on the left, two atoms on the right

The macrostate description contains inherently less information than the microstate description, but the macrostate
description accurately reflects the number of recognizable

Table 6.1

How the microstates map to the macrostates

Microstate	With labels	Macrostate	Without labels	W
1	Atom 1 on the left, atom 2 on the left	1	Two atoms on the left, zero atoms on the right	1
2	Atom 1 on the left, atom 2 on the right	2	One atom on the left, one atom on the right	2
3	Atom 1 on the right, atom 2 on the left			
4	Atom 1 on the right, atom 2 on the right	3	Zero atoms on the left, two atoms on the right	1

outcomes (in a world where atoms are packaged without labels). Table 6.1 shows how the microstates for the current system map to the corresponding macrostates.

The table also contains a column marked W, a promising sign. This quantity corresponds to the W that appears in Boltzmann's $S = k \log W$. Progress has apparently been made, but of what significance? The quantity W, called the multiplicity, refers to the number of microstates that contribute to a particular macrostate. W therefore emerges as a result of the key step, which is the lumping together of microstates. This step, which could easily go unnoticed, represents the decisive act by which Boltzmann revealed the true meaning of entropy. Entropy would not exist were it not for the lumping together of microstates. And why again did we lump together microstates? We lumped together microstates by necessity, because two configurations were indistinguishable. But we would likely find good reason to construct macrostates even if it were not necessary, even if we could observe the unobservable. Why? Because we rarely need to know the position of atom 1, the position of atom 2, and so on. The microstate description provides us with more information than what we typically need. We get along just fine without knowing where to find atom 1. What we really need to know is how the atoms are distributed across the room. We need to know that it's twice as likely that the atoms are evenly distributed, as in macrostate 2, than it is that both atoms are on the left, as in macrostate 1. Macrostate 2 is twice as likely as macrostate 1 because twice as many microstates contribute to it. Lumping together microstates would be useful even if it were not necessary.

What then might we say about our admittedly crazy question? Are we prepared to make sense of this auditorium whose occupants show such a strong predilection for the left-hand side? Not yet. Not with just two atoms. In a room with just two atoms, we'll routinely find both atoms on the left. We need to somehow restore common sense to ensure that crazy scenarios are in fact crazy. And for that, we need big numbers.

7 Not Quite Big Numbers

We need big numbers. But how big? Let's take a small step toward big numbers: go from two atoms to four. In a room filled with four atoms, there are sixteen possible microstates. We've not yet reached big numbers, but we already need an accounting system that will enumerate all of the ways that four atoms can be assigned to either the left- or the right-hand side of the room. One way to visualize this task is to treat these assignments as the branch points in a tree (figure 7.1).

There are two possibilities at the first branch point: atom 1 can go on the left or the right. This first assignment then leads to another series of branch points for atom 2, and another for atom 3, and another for atom 4. This strategy leads to four binary sets of branch points, giving $2 \times 2 \times 2 \times 2 = 16$ possible microstates, each represented by one of the terminal branches in the tree. If atoms 1, 2, 3, and 4 are consecutively assigned to the left, we arrive at the terminal branch on the far left. But this microstate represents only one way in which we might weave our way through the tree. There are fifteen other paths that are just as likely.

The microstates, however, do not tell the whole story. We need to throw away the labels and create the macrostates that will lead to entropy. Imagine then that you turn your back

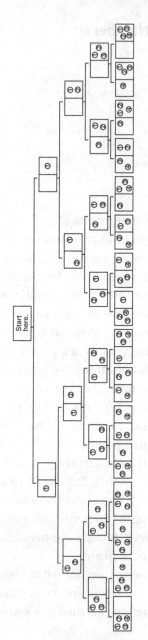

Figure 7.1
Tree diagram for four atoms

on the system, allow the system to sample one of the sixteen microstates that are available to it, and then turn back around. What would you see? In the absence of any labels, you would see one of five possible macrostates:

1. Four atoms on the left, zero atoms on the right
2. Three atoms on the left, one atom on the right
3. Two atoms on the left, two atoms on the right
4. One atom on the left, three atoms on the right
5. Zero atoms on the left, four atoms on the right

Now let us rely on intuition, a surprisingly poor guide throughout much of modern physics, but something that will serve us well here. Let's say that when you turn back around, you observe macrostate 1. You might be a bit surprised to see that all four atoms are on the left-hand side of the room. And if you're a curious sort of person, which you undoubtedly are (consider the fact that you're reading a book about a nineteenth-century physicist who suffered from a pretty bad case of anxiety while studying a thing called entropy), then you might be inclined to repeat the experiment. So let's say that you repeat the experiment, and next time around, you observe macrostate 2 (or 3 or 4 or 5). You now know that the system is not for some reason stuck in macrostate 1 and therefore has the ability to change. And if the system has the ability to change, you could continue to repeat the experiment until some statistics emerge. In doing so, you would eventually find that macrostate 1, despite its initial showing, occurs rather infrequently, as does macrostate 5. Macrostates 2 and 4 are more likely, but not as likely as macrostate 3, which is observed most frequently. These findings can be made quantitative by mapping the sixteen available microstates to the five macrostates (figure 7.2).

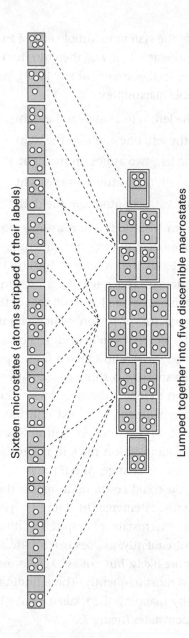

Figure 7.2

Mapping the microstates to the macrostates

Sixteen microstates (atoms stripped of their labels)

Lumped together into five discernible macrostates

The likelihood of a particular macrostate is given by the number of microstates that contribute to it. It is given by the multiplicity W. We know that macrostate 1 is unlikely because it includes only one microstate. There is only one way in which all four atoms can be arranged on the left-hand side of the room. Only one path through the tree leads to that outcome. Macrostate 2 is more likely because there are four different ways to arrange three atoms on the left and one atom on the right. Simply pick the atom that you wish to place on the right. If you pick atom 1 to be on the right, then atoms 2, 3, and 4 must be on the left. But that is only one of four possibilities. You could just as easily pick atom 2 to be the sole representative on the right, or atom 3, or atom 4. All four of these configurations look the same, and when they are lumped together, they constitute macrostate 2. This macrostate is not insignificant because it accounts for four of the sixteen available microstates (25 percent of all possible configurations). The only macrostate that includes a larger number of configurations is macrostate 3 with its six configurations, all of which display two atoms on the left and two on the right. We might have hoped that this most likely macrostate would exert a more dominant effect on the landscape of available configurations and that the atoms would exhibit an even stronger tendency to distribute themselves equally on both sides of the room. But the evidence so far gives us reason to worry because six is not sixteen. Even this most likely macrostate is by itself outnumbered, leaving ample opportunity for the atoms to concentrate on one side of the room or the other.

Imagine what havoc this scenario would wreak upon our lives if it were a cause for legitimate concern. Before installing a treadmill, you would need to consider whether some part of

the room was preferable, if some corner provided a particularly
plentiful supply of air molecules for your future workouts. You
can disregard these considerations because they are in fact not
a legitimate concern. You can install the treadmill on the left
or on the right or in the middle. Or if installing a treadmill is
the least of your worries, you can rest assured that the room in
which you now sit (comfortably, I hope) harbors no unpleasant surprises for you when you get up. You'll be able to walk
around the room in any direction that you like, and equal
numbers of air molecules will greet you on the left and on the
right. Why? Because your room has more than four atoms.
We've not yet reached big numbers.

8 Big Enough Numbers

It's time to take the leap from four atoms to a roomful of atoms. We don't need an enormous room like an auditorium. A small, boxy room—12 feet × 12 feet × 10 feet—will work. (Apologies to the metrically inclined.) At room temperature and atmospheric pressure, a room of this size contains around 1,000,000,000,000,000,000,000,000,000 (a 1 followed by twenty-seven 0s) gas particles. This number is obviously big but not so big as to be nameless. The name is supposedly one octillion, which sounds made up, but is in fact no more made up than the name of any other number. We are probably obliged here by popular science convention to channel our inner Carl Sagan and provide some fact or figure that conveys the jaw-dropping enormity of one octillion. Is it comparable to the number of cells in the human body? To the number of grains of sand on all the world's beaches? To the number of stars in the observable universe? All of these quantities reach high into the *-illions*, but none of them as high as one octillion. Perhaps the last of these comparisons is the most illuminating—that a modest-size room contains more gas particles than there are stars in the universe. Sagan would have said it better—something about a universe teeming with stars (billions and billions of them), with worlds of different scales nested one inside the other. Sagan helped us to feel at home in the universe—as if we might actually belong here, not trapped on some rapidly spinning rock, along with a bunch of mostly crazy strangers, destined to go around and around an unremarkable star amid a mostly empty space. He was a reassuring tour guide in a universe where we play a seemingly inconsequential role. He offered us awe and wonder when we might otherwise have succumbed to a sense of insignificance. The cosmos seemed a more welcoming place because he shared it with us.

Let us return to the task at hand.

The gas particles in our room are a mixture of elements and compounds, some of them atoms and others of them molecules. For convenience, we'll ignore these differences and refer to all of the particles as atoms, so we have 1 octillion atoms to consider. The exercise is the same as before. Each atom can go on the left- or the right-hand side of the room. The total number of ways to arrange the atoms—the number of microstates—is then $2 \times 2 \times 2 \times \ldots$ and continue to multiply by 2 until we get to 1 octillion 2s. Many readers undoubtedly know something (or many somethings) about exponents and recognize that the number of microstates can be written as $2^{\text{octillion}}$.

Now 1 octillion (the number of atoms) and $2^{\text{octillion}}$ (the number of microstates) might appear to be similar quantities because both include this really big number called an octillion. But the use of the exponent elevates $2^{\text{octillion}}$ to heights that render 1 octillion practically insubstantial. In fact, if we were to draw a number line that provided enough room for the number $2^{\text{octillion}}$, the number 1 octillion would not even register. One octillion would fall not a hair's breadth from the zero mark. It would land pretty much on the origin of the number line. Recall that we can actually write out the number 1 octillion. It's just a 1 followed by twenty-seven 0s. But any attempt to write out the number $2^{octillion}$ would be a waste of time. By the time that you multiply out all of those 2s, you would have a number that defies decimal notation. You could start with a 1 (easy enough), but you would need to follow the 1 with over 300 septillion 0s. We could define a septillion for you, but you might be satisfied simply in knowing that 300 septillion 0s give a bigger number than twenty-seven 0s. Behold the power of the exponent. It has led to territory that defies facts and figures. One potentially useful comparison is the number of atoms in the observable universe. That number is estimated to be a measly 1 followed by eighty 0s. And again, a 1 followed by 300 septillion 0s is a bigger number than a 1 followed by eighty 0s. We cannot exaggerate the magnitude of $2^{octillion}$.

How is it that the number of atoms in a room gives rise to a number of microstates that vastly exceeds the total number of atoms in the universe? A tree diagram might help. We previously constructed a tree diagram for a system of four atoms. The tree in that case branched at four different levels to give sixteen terminal branch points, one for each of the system's sixteen microstates. Now imagine a tree that seems as if it might never stop branching. Each level is followed by an even more densely packed level beneath it, initially creating a pleasant cascading effect, but quickly reaching a point of saturation wherein all structure is lost. And the tree just continues branching. It has not four, but 1 octillion levels for which it must account. Its terminal branches, if we could see them, would number $2^{octillion}$ or, if you prefer, a 1 followed by 300 septillion 0s. This procedure, in which branches give rise to branches (which give rise to branches), is achieved mathematically by use of the exponent. And when given the chance, the exponent allows even the number 2 to blow up.

We must, however, remember that our task is not simply a matter of generating microstates. Microstates lead to macrostates, which lead to entropy. The microstate that initiated this conversation, the one in which all 1 octillion atoms had drifted over to the left-hand side of a room, is a macrostate. But very few microstates are also macrostates. In fact, only two of the $2^{\text{octillion}}$ microstates in this example are macrostates. One of these microstates assigns the atoms uniformly to the left, while the other assigns the atoms uniformly to the right. What makes either of those two configurations particularly special? Consider the path that leads to the uniformly left-leaning configuration. We can trace this particular path by asking the question *Left or right?* 1 octillion times. If the answer is *left* for atom 1, *left* for atom 2, *left* for atom 3, and so and so on, then the path hugs the extreme left-hand side of the tree diagram and ultimately arrives at a configuration that cannot be achieved by any other network of branches. It's an outcome that requires an unwavering monotony. Given 1 octillion opportunities to answer differently, the answer is always the same. This path leads to a microstate that stands alone, a microstate that is its own macrostate.

Contrast this scenario with one that is much more likely—equal numbers of atoms on the left and on the right. Many roads lead to that destination. We can wander along the tree diagram, going left and right and not worrying too much about the details of the trip, so long as the number of left turns ends up being equal to the number of right turns. We can assign the atoms *left, right, left, right, left, right, …* and continue on like that until all 1 octillion atoms have been assigned. Or we can assign the first half-octillion atoms to the left and remaining half-octillion atoms to the right. Or we can find some more jumbled-up sequence of assignments that satisfies our one condition—that the two sides acquire equal numbers of atoms in the end. Impartiality of this sort runs rampant throughout the tree diagram; many paths lead to microstates that would be lumped into this macrostate. And the likelihood of a macrostate grows with the number of its constituent microstates. The likelihood, therefore, of finding equal numbers of atoms on both sides of the room is relatively high because the macrostate that describes this scenario is one that accounts for a significant percentage of the available microstates. And the probability of finding all of the atoms on the left-hand side of the room is especially low because this configuration requires a very specific set of circumstances. Of the $2^{\text{octillion}}$ microstates that are available to the system, only one represents the room that is completely lopsided on the left. And no one would place a bet if the odds of winning were 1 in $2^{\text{octillion}}$. (A much safer bet would be to pick one supposedly lucky atom from all of the other unlucky atoms in the universe.) We have to accept the fact that some things are possible but so unlikely as to be effectively impossible. We'll never observe the completely lopsided room. I use the word *never* here despite the fact that the outcome is not absolutely forbidden, and I do so with a clear conscience because it would be foolish to await all possible outcomes. We are sensible people who live for a finite time. The chances might not be zero, but they are 1 in $2^{\text{octillion}}$. And at some point, close-enough-to-zero is effectively zero. On the list of all things to safely ignore, we can (and should) include the completely lopsided room. We are as close to certainty here as is permitted in science. The admittedly crazy question is in fact crazy.

9   And Everything In Between

We've established two things so far: (1) It's very unlikely that all of the air will migrate to one side of a room, and (2) it's much more likely that the air will distribute itself evenly, with equal numbers of atoms on both sides. (The theme of this book is apparently to expect the expected.) But we've neglected all of the possibilities that lie somewhere in between these two extremes. For example, do we really expect to observe *exactly* equal numbers of atoms on both sides? Such a condition imposes an improbable symmetry that itself seems too stringent. Is not a slight excess of atoms on one side of the room inevitable?

In order to answer these questions, we need a way to account for any irregularities. But in a microscopically messy room with 1 octillion atoms, how would we even know if a few too many atoms had drifted momentarily to the left? At what point would our senses alert us that the scales had somehow tipped? The fact that alarm bells of this sort fail to ring suggests that any irregularities are small, but let's not rely too heavily on our senses. We don't want our conclusions to rest on the sensitivity (or insensitivity) of some air-sampling center of our brains. Our powers of abstraction will be much greater here than the powers of our senses. We can count microstates, and microstates allow us to calculate the likelihood of the irregularities.

It might help if we take a step back, returning to our four-atom system with its sixteen microstates. (Apologies to those of you who are now drunk on the thought of big numbers; a return to four atoms must seem a real letdown.) But rather than carelessly lumping together microstates as we did before, let's stack the microstates neatly, one on top of the other, so as to build a column of microstates, one for each macrostate (figure 9.1).

Then let's arrange those columns in some meaningful order. For example, each macrostate can be uniquely identified by the number of atoms that are present on the right-hand side of the room. So we can arrange the macrostates in increasing order from zero atoms on the right to four, which gives us something called a histogram (figure 9.2).

This particular histogram tells us the likelihood of each of the macrostates. Each macrostate is represented by a column whose height is proportional to the number of microstates that contribute to it. The height gives the multiplicity W. Common macrostates with lots of microstates rise up high, and rare macrostates with relatively few microstates lie down low.

When viewed in its entirety, the histogram tells us how the microstates are distributed across the available macrostates. It tells us where the microstates have been binned. (We're binning now rather than lumping.) Sometimes the binning process piles microstate on top of microstate, with a few columns rising high above the others. The distribution in that case is narrow and focuses attention sharply on the few macrostates that matter. But at other times, the microstates spread out more equitably, so that the histogram is more

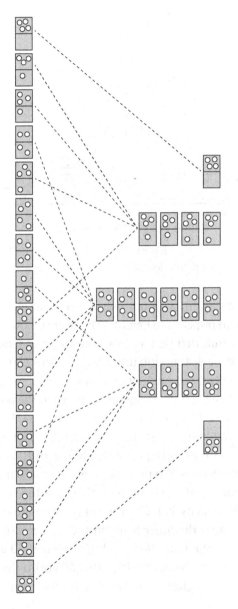

Figure 9.1
Binning of microstates

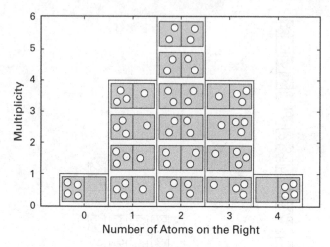

Figure 9.2
Construction of a histogram

mesa than Matterhorn, with not one or a few but many columns that rise to respectable heights. The distribution then is
broad and demands that we take lots of macrostates seriously.

The shape of the distribution ultimately depends on the size
of the system. When the number of atoms is small, the distribution is broad. But it narrows significantly when the number
of atoms increases. This trend is illustrated in figure 9.3. The
histogram on top corresponds to a system of 10 atoms, and
the one on the bottom corresponds to a system of 100 atoms.

In each of these cases, the most likely macrostate is the one
with equal numbers of atoms on both sides. For the 10-atom
system, the multiplicity is highest when 5 atoms are on the
right (implying that the other 5 are on the left). And for the
100-atom system, the multiplicity is highest when 50 atoms
are one the right (implying that the other 50 are on the left).
The two cases differ significantly, however, in the distribution

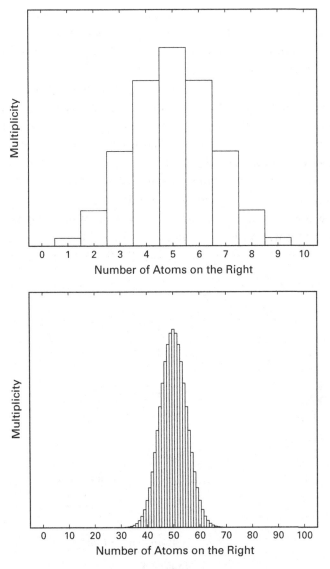

Figure 9.3
Histograms for systems with 10 (top) and
100 (bottom) atoms

of microstates about these most likely outcomes. In the case of the 10-atom system, all but the most extreme macrostates make significant contributions to the histogram. We can safely ignore the macrostates in which 0 or 10 atoms are on the right, but all of the other macrostates are significant in that they are represented by columns that rise noticeably above the baseline. We can conclude, therefore, that if the system has only 10 atoms, we should not be too surprised to find it configured with only 3 atoms on the right. This outcome is not the most likely, but it's not ridiculously improbable either. Small numbers allow large irregularities. Big numbers, however, are far less tolerant. Given the fact that a 10-atom system might be observed with as few as 3 atoms on the right, we might suspect that a 100-atom system is sometimes observed with only

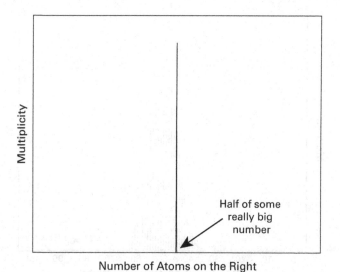

Figure 9.4
Histogram for a system with a much larger number of atoms

30 atoms on the right. We simply scale both numbers by a common factor of 10. Ten goes to 100, and 3 goes to 30. Does the conclusion then not hold?

No, it does not, and the figure tells the story. At the mark where 30 atoms are on the right, we do not see a column. And we do not see a column because the column that resides there is very tiny. It cannot compete with the columns that represent the more likely macrostates. It is not even a blip on their baseline. This distribution is owned by the macrostates in the middle. So we might expect a 100-atom system to be configured with 45 or 55 atoms on the right but not 30. Forget about 20, and 10 is almost unimaginable. Bigger numbers impose a greater conformity on the system.

And 100 atoms does not even make for a very big system. As the number of atoms increases toward the –*illions*, the distribution narrows even further, and the histogram stretches upward. Irregularities persist, but they become smaller and smaller compared to the size of the system. They go unnoticed by the senses and resist depiction, and the descent from Mount Histogram occurs so precipitously—from most likely, to likely, to maybe, to forget about it—that the structure resembles nothing but a needle in the sky (figure 9.4).

The take-home message is that in a room with an -*illion* atoms, the irregularities end up looking pretty regular.

10 A Case for Anxiety

The patient is long dead. The opportunity for a psychiatric consult is forever lost. The surviving medical record is incomplete and a product of its time. We're in no position to speculate, and yet we wonder. Can conventional wisdom be trusted? Have we inherited an accurate diagnosis? Was Boltzmann really bipolar?

There's ample evidence that he was depressed, some of which was the result of normal aging. Senses diminish and novelty wanes over the course of a long-lived life, and old age often brings with it real physical pain. Boltzmann suffered all of these common travails. His eyes were a particular problem. It's sometimes noted that he failed to recognize the remarkable discoveries that were made at the start of the twentieth century. He seemed hardly aware, for example, that Planck and Einstein had made extraordinary leaps forward, with his ideas serving as their primary source of inspiration. But consider the challenge of remaining up-to-date in a rapidly changing field, when every time that you wanted to read a paper, you needed to find an assistant or family member who would read that paper to you. His failing eyes also stole from him the ability to read music. Toward the end of his life, friends remember him hunched over the piano jostling multiple pairs of eyeglasses.[1] The piano had always been a place where he could convene with others, and evenings at the piano with family and friends had always been a welcome part of his routine. But as his eyesight worsened, those gatherings became less frequent, which left him feeling isolated and increasingly depressed.

A diagnosis of bipolar disorder, however, requires that the patient experience manic highs along with depressive lows. And here we diverge from the traditional story. We know that Boltzmann was excitable and sometimes irritable. We also know that he liked to work. He often started his day before dawn and did not always end his day at a regularly scheduled hour. He was highly productive. But it seems an exaggeration to say that he was manic. A disclaimer here: this view is offered not by a mental health professional. I am a chemist who offers none of the credentials that are needed in order to make a definitive diagnosis of anyone living or dead. But still it seems that Boltzmann's manic episodes are more assumed than they are documented. History found it fitting to diagnose him with bipolar disorder, and it became part of his story. And once a good story has been established, there is often little incentive to revise it. But here I'll tell a different story. We'll take a leap down a different rabbit hole in the hope that we might land some place dark but not altogether uninviting—some place in need of exploring.

I'll revise the traditional story by reclaiming the original diagnosis of neurasthenia, which Boltzmann's contemporaries assigned to him. This disorder lacks the specificity of a modern diagnosis. It was not a simple mood disorder, although it included symptoms that today would serve as check boxes for depression. It was a know-it-when-you-see-it catchall diagnosis that provided a label for problems that were not yet even vaguely understood at the time.[2] His caretakers may have had good intentions, but they had no reliable diagnostic criteria. They knew nothing about neurotransmitters, and they offered few, if any, effective therapies (be they psycho- or pharmaco-). If he were alive today, Boltzmann's prognosis might actually be good, but he lived at a time that was ill prepared to treat him.

Today he would at least be provided with a reliable diagnosis. And I suspect that the primary diagnosis would be something too little appreciated thus far in the conversation. Today he would most likely be diagnosed with an anxiety disorder. Was he depressed? Yes. Was he excitable, irritable, and maybe even hypomanic? Fair enough. But each of these symptoms was likely not so much a cause as it was an effect. The underlying cause of his instability was likely anxiety. Boltzmann was a nineteenth-century victim of a twenty-first-century disease.

11   One of the Many Benefits of Modernity

Did early man (or woman, or child) suffer from anything that would today be classified as anxiety? It would seem that early humans had every reason to be anxious. Their predators were not conveniently locked up in zoos for their amusement. Their food did not predictably await their return to the supermarket. Their world in fact was scary and hard in ways that ours is not. (By *ours*, I mean those fortunate enough to enjoy the luxury of reading a book about Ludwig Boltzmann.) We might assume that clinical anxiety was rampant then, that our earliest ancestors waited fretfully outside the cave of some wise, old shaman, hoping that the man might dispense some voodoo-magic peace of mind. We might imagine those first patients to be Boltzmann's psychological forebears.

Yet our earliest ancestors would surely regard Boltzmann as a strange creature and his problems as inconceivable. Early humans had real problems that demanded their immediate attention. Their fear was justified and served a useful purpose because it drove them to act (at a time when inaction was often inadvisable). Boltzmann, however, never went hungry a day in his life. He somehow managed to survive on a steady diet of pork, cabbage, potatoes, and beer. And his natural enemies were not lions and tigers and bears but privileged professors stuffed into starched shirts from which protruded well-coiffed beards. He faced real challenges, but nothing that threatened his survival and nothing that should have kept him up at night. He was a modern man with modern problems, his disorder one that emerges only after primal obligations have been met. Workplace feuds and intellectual disputes become objects of concern only when the belly is full and the vultures lurk elsewhere.

But Boltzmann's problem was not altogether unrelated to those ancestral fears. Anxiety is what remains of fear after basic needs have already been met. It's a vestigial fear, arising from some old part of the brain, some part that's not yet had time to recognize the recent changes in its environment. (By *recent*, I mean the past few hundred years.) This old brain is like a computer that seldom checks for updates, its hardware designed for yesterday's applications. It sometimes fails to make useful distinctions and can leave us feeling fearful when we have every reason to be carefree. We live in a brave new world, but we're equipped with a frightened old brain, and anxiety is a price we pay for living in this newly found, particularly auspicious sliver of the human time line.

When faced with too few targets, our psychological defenses have a tendency to fill in the blanks. We find reasons to be afraid. We find fears that are incommensurate with actual threats, fears that are tied to things that might not even happen, things that don't even need to happen. Anxiety is often irrational, and therein lies its most maddening feature: it turns otherwise rational people into lesser versions of themselves. Boltzmann was a rational man prior to his illness. Actually he was more than rational: he exemplified the power of human reason. But anxiety has a way of crowding out higher, more productive functions. It expands across the space that's provided to it, often leaving little room for anything else. And so the brain that gave us $S = k \log W$ was recommissioned for useless other purposes that hurt its proprietor and hindered a great intellect.

12   And Yet It Moves

I noted at the start that atoms have both positions and velocities and accounted for the positions, albeit roughly, with the designations *left* and *right*. I also introduced the concepts of microstates and macrostates and multiplicities and promised that all of these things have something to do with entropy. I mentioned, in particular, that the multiplicity is designated by the letter W, which is the same W that appears in $S = k$ log W. We might understandably want to now lunge for the goal—to make quick sense of that log function, to multiply by the constant k and grab for the elusive entropy S. But then we would not know the role that's played by the velocities, and we would miss a significant part of the story. So let's resist the urge to claim victory prematurely. Let's breathe deep and think deeper and release ourselves from the pull of persistent distractions—emails and tweets and status updates—all of which never go away and always seem in need of constant checking. The second law offers no such instant gratification. It provides instead an opportunity to swim in deep waters, to go where the heart slows and the mind wanders and modern trappings (that no doubt contribute to our anxiety) sometimes fade away.

So where do we go next? The atoms move and so have velocities. We'll not focus here on the directions in which the atoms move. A more ambitious treatment of the second law might specify that a particular atom is moving left or right, up or down, forward or backward, or some combination thereof. But a book that aims to be simple provides only the most essential parts of the story. We'll disregard the directions in which the atoms move and focus only on their speeds. The speeds tell us how fast or slow the atoms are moving. For accounting purposes, we might also prefer, as Boltzmann sometimes did, to monitor a related quantity called the kinetic energy. Discussions of kinetic energy generally start with the revelation that the word *kinetic* originates from the Greek word for *motion*. The reader is then alerted to the fact that the concept must be important because the Greeks had a word for it, and the author feels that he or she has done due diligence by following the conventional script. But in this particular case, the word's origin actually provides us with some insight. Kinetic energy is in fact the energy due to an object's motion. Fast-moving objects have lots of kinetic energy, and slow-moving objects have relatively little. But how much is lots? And how little is little? And what are the consequences? Let's turn, as is our habit, to a concrete example. Let's go back to the beginning.

In the beginning, there were two atoms. For the sake of this example, let's say that the kinetic energy of each atom can be 0, 1, or 2. You might ask (especially if you've survived some formal scientific training) that we assign units to these values. You might want to know if the energy is measured in joules or calories or some other agreed-on base unit. Our goal here, however, is not so much to calculate as it is to count, so there's no real need right now to ground these values in a standard system of measurement. The important thing to recognize is that each atom can have some kinetic energy, and that energy can be 0, 1, or 2.

Next, we need to count up all of the ways in which energy can be distributed across the two atoms. The tree diagram in this case (figure 12.1) has only two levels because the system has only two atoms. But each branch gives rise to three new branches because there are three possible energy values.

The energy of atom 1 is assigned at the first level. If the energy is 0, we'll represent that designation mathematically as $\varepsilon = 0$; if the energy is 1, then $\varepsilon = 1$; and if the energy is 2, then $\varepsilon = 2$. That gives three possibilities for atom 1. Each of these three possibilities is then combined with one of three possibilities for atom 2. The tree diagram therefore has $3 \times 3 = 3^2 = 9$ terminal branches. The branch on the far left represents the case in which both atoms are assigned 0 energy, which would be an especially lethargic collection of atoms, while the branch on the far right represents a much more spirited collection of atoms, both having an energy of 2. And in between these two extreme cases lie seven other possibilities.

Having traced out the tree diagram to each of its terminal branches, you might naturally think to call these different possibilities microstates and to lump the microstates together into macrostates. But an additional step is needed when assigning energies. We need to consider the total energy that's present in each of the nine cases. The case on the far left has a total energy of 0 (two atoms, each with zero energy: $2 \times 0 = 0$), while the case on the far right has a total energy of 4 (two atoms, each with an energy of two: $2 \times 2 = 4$). And there we have a problem. The total energy needs to be more clearly defined; it can't be 0 one moment and then 4 the next—not for an isolated system. Atoms don't sneak in or out of an

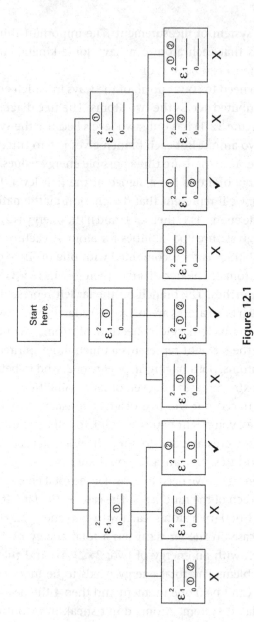

Figure 12.1
Tree diagram for the assignment of energies

isolated system, and neither does energy. These quantities are constant in an isolated system. We must be mindful, therefore, to conserve them. We must respect the conservation of mass as well as the conservation of energy, two principles that underlie much of physics. We need to pick some value for the total energy and enforce it.

Given the fact that the two extreme possibilities are 0 and 4, a total energy of 2 seems a reasonable compromise between the sluggish and the wildly energetic. And having selected that value, we can then trim the tree. For this particular example, the tree is trimmed of all but three branches because only three of the nine branches have a total energy of 2. These three branches are indicated in figure 12.1 by check marks, and the remaining branches (i.e., those in need of trimming) are indicated by Xs. And now we have our microstates

1. Atom 1 has $\varepsilon = 0$; atom 2 has $\varepsilon = 2$.
2. Atom 1 has $\varepsilon = 1$; atom 2 has $\varepsilon = 1$.
3. Atom 1 has $\varepsilon = 2$; atom 2 has $\varepsilon = 0$.

The total energy in this case acts as a constraint, an additional criterion that needs to be met in order to qualify as a relevant microstate. We face constraints every day, often to our dismay. You could, for example, think of all sorts of ways to spend a paycheck. You could spend some money on food, some on housing, some on entertainment, and so on, and each of these amounts could vary depending on your priorities. But it makes no sense to think of all the ways that you might spend a paycheck until you first know its total value. The paycheck is a constraint that limits your list of possibilities. You can't shift money from the food budget to the entertainment budget if the entertainment budget already exceeds the total amount of the paycheck. Similarly, the total energy constrains the behavior of the two atoms. The atoms can't be more or less energetic (collectively) than what's dictated by the constraint.

Having accepted the fact that energy and paychecks are finite resources, we can then construct the macrostates. The drill is the same as before, although there's little lumping to be performed with only three microstates. We willingly turn our backs on the system, await the removal of any labels, and then turn back around. And what do we see? If confronted by microstate 1 or microstate 3, we see a particularly slow atom (one with 0 energy) and a particularly fast atom (another with an energy of 2). We would not know, however, which of the two atoms, 1 or 2, is moving slowly and which is moving fast (because atoms of the same type do not have labels or name tags or any other distinguishing feature). We have no choice but to lump these two microstates together. We'll call the result macrostate 1 and assign it a multiplicity of 2. The only other possibility, on turning back around, is microstate 2, in which the atoms appear to be neither especially fast nor especially slow. The energy of each atom is an intermediate

Table 12.1

How the microstates map to the macrostates

Microstate	With labels	Macrostate	Without labels	W
1	Atom 1 has $\varepsilon=0$; atom 2 has $\varepsilon=2$.	1	One atom with $\varepsilon=0$, one atom with $\varepsilon=2$	2
3	Atom 1 has $\varepsilon=2$; atom 2 has $\varepsilon=0$.			
2	Atom 1 has $\varepsilon=1$; atom 2 has $\varepsilon=1$.	2	Two atoms with $\varepsilon=1$	1

value of 1. This microstate is unique and can be identified with or without labels. It is its own macrostate and its multiplicity is 1. We'll call it macrostate 2.

And that's it, at least for two atoms. But if kinetic energy is to provide us with any real insight, then we'll need a bigger system.

13 Follow the Leader

We don't need to invoke ridiculously big numbers in our treatment of kinetic energy. Boltzmann needed no more than seven atoms. He made available to those seven atoms a total energy of 7 and thereby revealed the essential trend. His findings appeared in a landmark paper that hinted that S might equal something like $k \log W$.[1] Consider, he suggested, a system of seven atoms. Each atom can have an energy of 0, 1, 2, 3, 4, 5, 6, or 7, but the total energy must be 7. An atom therefore might have an energy of 7, but only if the remaining six atoms all have 0 energy. Or if the energy is distributed more equitably, then each atom could have an energy of 1. Or the energies might be distributed in some more complicated fashion. There are many possibilities to consider. The question is whether Mother Nature distributes her resources fairly. She might, in fact, be shamefully tolerant of inequality (not unlike the not-so-invisible hand of the modern market).

In order to reveal her essential character, we need to start counting, and seven atoms seems a manageable number. But a system of seven atoms already gives many more microstates than can be conveniently counted on a printed page, and the corresponding tree diagram is even worse. The tree, if we were to sketch it, has seven levels (for seven atoms), with each

branch giving rise to eight new branches (because there are
eight possible energy values). That tangled mess of a tree has
$8 \times 8 \times 8 \times 8 \times 8 \times 8 \times 8 = 8^7 = 2,097,152$ terminal branches. Most
of those branches, however, give a total energy equal to some-
thing other than 7. In fact, only 1,716 branches are consistent
with the total energy constraint. But 1,716 microstates are still
too many to list on page. Fortunately, the number of macrostates is not too unwieldy. After lumping, the 1,716 microstates
consolidate into a mere 15 macrostates (table 13.1).

One can quickly check that each macrostate has a total
energy of 7. A bit trickier task is to calculate the multiplicity W. An especially well-informed (or motivated) reader
knows (or could find) the multiplicity by using something
called a factorial function. But one doesn't need the factorial function for a few of the more straightforward macrostates.
Consider, for example, macrostate 1, to which we alluded earlier. Macrostate 1 corresponds to a despotic world in which all
of the energy resides with one atom. That lucky atom could
be atom 1, or atom 2, or atom 3, on down the line (one atom
to rule them all), but we don't know which atom. The system
would look the same to us regardless of which of the seven
atoms was the energetic king. So we have a macrostate with
seven microstates, a multiplicity $W = 7$, and nothing much to
worry about because seven is a small part of 1,716. Mother
Nature apparently does not favor a despot. But she has even
less use for the egalitarian world in which every atom has
an energy of 1. That world is one in which all of the atoms
rise (or fall, depending on your politics) to the same level
and no one atom takes more than its fair share. That macrostate has a multiplicity of $W = 1$ because there's only one
microstate that contributes to it. It represents a unique path
down the tree diagram—a path in which every atom, when

Table 13.1
Macrostates for seven atoms with a total energy of seven

Macrostate	Energy distribution	Total energy check (Make sure that the answer is seven.)	W
1	Six atoms with $\varepsilon = 0$, one atom with $\varepsilon = 7$	$(6 \times 0) + (1 \times 7) = 7$	7
2	Five atoms with $\varepsilon = 0$, one atom with $\varepsilon = 1$, one atom with $\varepsilon = 6$	$(5 \times 0) + (1 \times 1) + (1 \times 6) = 7$	42
3	Five atoms with $\varepsilon = 0$, one atom with $\varepsilon = 2$, one atom with $\varepsilon = 5$	$(5 \times 0) + (1 \times 2) + (1 \times 5) = 7$	42
4	Five atoms with $\varepsilon = 0$, one atom with $\varepsilon = 3$, one atom with $\varepsilon = 4$	$(5 \times 0) + (1 \times 3) + (1 \times 4) = 7$	42
5	Four atoms with $\varepsilon = 0$, two atoms with $\varepsilon = 1$, one atom with $\varepsilon = 5$	$(4 \times 0) + (2 \times 1) + (1 \times 5) = 7$	105
6	Four atoms with $\varepsilon = 0$, one atom with $\varepsilon = 1$, one atom with $\varepsilon = 2$, one atom with $\varepsilon = 4$	$(4 \times 0) + (1 \times 1) + (1 \times 2) + (1 \times 4) = 7$	210
7	Four atoms with $\varepsilon = 0$, one atom with $\varepsilon = 1$, two atoms with $\varepsilon = 3$	$(4 \times 0) + (1 \times 1) + (2 \times 3) = 7$	105
8	Four atoms with $\varepsilon = 0$, two atoms with $\varepsilon = 2$, one atom with $\varepsilon = 3$	$(4 \times 0) + (2 \times 2) + (1 \times 3) = 7$	105
9	Three atoms with $\varepsilon = 0$, three atoms with $\varepsilon = 1$, one atom with $\varepsilon = 4$	$(3 \times 0) + (3 \times 1) + (1 \times 4) = 7$	140

(continued)

Table 13.1 (continued)

Macrostate	Energy distribution	Total energy check (Make sure that the answer is seven.)	W
10	Three atoms with $\varepsilon = 0$, two atoms with $\varepsilon = 1$, one atom with $\varepsilon = 2$, one atom with $\varepsilon = 3$	$(3 \times 0) + (2 \times 1) + (1 \times 2) + (1 \times 3) = 7$	420
11	Three atoms with $\varepsilon = 0$, one atom with $\varepsilon = 1$, three atoms with $\varepsilon = 2$	$(3 \times 0) + (1 \times 1) + (3 \times 2) = 7$	140
12	Two atoms with $\varepsilon = 0$, four atoms with $\varepsilon = 1$, one atom with $\varepsilon = 3$	$(2 \times 0) + (4 \times 1) + (1 \times 3) = 7$	105
13	Two atoms with $\varepsilon = 0$, three atoms with $\varepsilon = 1$, two atoms with $\varepsilon = 2$	$(2 \times 0) + (3 \times 1) + (2 \times 2) = 7$	210
14	One atom with $\varepsilon = 0$, five atoms with $\varepsilon = 1$, one atom with $\varepsilon = 2$	$(1 \times 0) + (5 \times 1) + (1 \times 2) = 7$	42
15	Seven atoms with $\varepsilon = 1$	$7 \times 1 = 7$	1

given the opportunity to distinguish itself, elects instead to follow the lead of all of its predecessors and successors. This macrostate is not unlike the lopsided room that we previously relegated to the world of the crazy.

We therefore need to focus on the macrostate that comprises the largest number of microstates. If we accept the fundamental postulate—that all microstates are visited with equal likelihood—then the macrostate with the largest number of microstates will be observed most frequently. And in this particular example, that particular macrostate is macrostate 10, which accounts for 420 of the 1,716 microstates. No other macrostate can claim such a significant fraction of the available landscape. Mother Nature's true character is now revealed: the most likely macrostate is one in which three atoms have 0 energy, two atoms have an energy of 1, one atom has an energy of 2, and one atom has an energy of 3. This energy distribution can be depicted as another sort of histogram (figure 13.1).

The histogram indicates that for the most likely macrostate, the number of atoms with a particular energy decreases as the energy increases. Most of the atoms have relatively

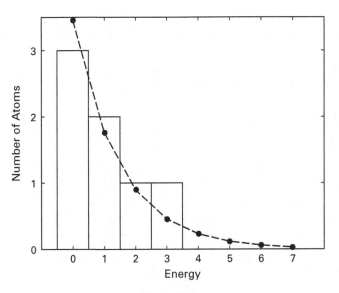

Figure 13.1
Histogram for the most likely macrostate

little energy (ε=0 or 1), a few atoms have a slight excess of energy (ε = 2 or 3), and none of the atoms has an energy greater than 3. This distribution is considered most likely because it represents the macrostate with the largest number of microstates (i.e., the macrostate with the highest multiplicity). It's a rough approximation of something called the Boltzmann distribution, which is represented by the curve that traces a descent down the histogram. The Boltzmann distribution is a prominent example of what is generally called an exponential distribution. It's the calculus-based solution to the problem of finding the most likely macrostate. Although our tools here have been much more basic—we've simply counted our

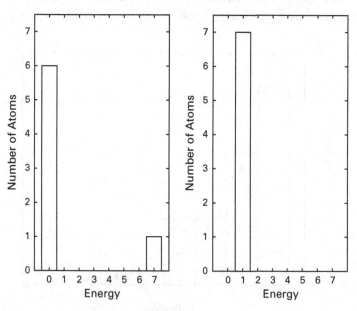

Figure 13.2
Histograms for two less likely macrostates

way to the most likely macrostate—we can now call this result the Boltzmann distribution.

The Boltzmann distribution contrasts sharply with the histograms for either of the two extreme scenarios that we previously considered.

The histogram on the left in figure 13.2 is sharply peaked at 0 energy, save for the one lone count that's attributed to the despot at $\varepsilon = 7$, while the histogram on the right is even more sharply peaked at an all-atoms-are-created-equal energy of 1. And we now know from the multiplicity values that the Boltzmann distribution is $420/7 = 60$ times more likely than the macrostate on the left and $420/1 = 420$ times more likely than the macrostate on the right. The Boltzmann distribution,

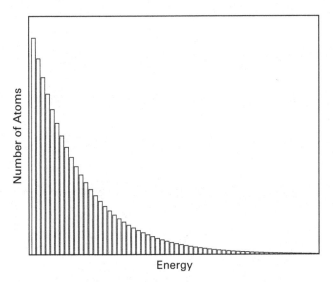

Figure 13.3

A more smoothly decreasing exponential function, characteristic of the Boltzmann distribution with its inexorable slide toward the baseline, limits most atoms to relatively low energies but allows a small number of atoms a modest surplus of energy. This trend is characteristic of the most likely macrostate. And that trend is already apparent with as few as seven atoms. Systems with more than seven atoms give more smoothly varying Boltzmann distributions, but the trend is the same as before, which allows the initiated (of which you are now a member) to recognize Mother Nature by one of her defining features.

14 The Night Before

It was getting late, and he was developing some bad habits. She walked down the hallway to the study and looked inside. There she found her husband huddled over his notes, his face scrunched up. She didn't knock. He didn't notice her. She knew his schedule. It included only an introductory physics lecture the next day, which he should have been able to give in his sleep. And sleep is what he really needed. Yet here he was, acting as if he still needed to prepare.

He had not always been this way. She had sat in on several of his classes before they were married, despite some protest.[1] Many of his colleagues (a real enlightened bunch) thought it unwise to allow a female student in the classroom (something about cooties maybe).[2] But Boltzmann had never stood in her way. His views on the subject were rather liberal, and she was persistent. She lobbied the administration until she found her way into his classroom. And then she lobbied him personally until she found her way into his life on a more permanent basis. What was his role in these developments? Something in between a clueless bystander and a willing accomplice to it all. But he was lucky that she had chosen him.

Many years later, however, their luck had run out. She had tied her fortunes to his, and he had ceased to be a wise investment. He had given thousands of lectures over the course of his career, presented his work hundreds of times before the Imperial Academy of Sciences, and traveled abroad to stand before august gatherings of supposedly learned men. But now he struggled to face a group of first-year college students. And he was getting worse, trapped in some self-destructive loop. Each day he would give his lectures and shortly after would start to worry about the next day. It was nonsense. Rather than enjoying the success that he had earned over the course of his career, he was worrying about tasks that he had already performed a thousand times before. He was disconnected from reality, and she had no idea how to help him.

"Ludwig, it's time to go to bed," she said. She sounded sympathetic. She was sympathetic. But she didn't understand. How could she understand? Boltzmann looked up, laid down his notes, and offered her something like a smile. He started to protest, grumbling that he still had some work to do but was almost finished. She sighed, gave her husband a goodnight kiss and a ten-minute deadline, and headed off to bed.

15   The Next Day

The professor shuffled in, shoulders hunched, eyes downward. His general appearance was suitably disheveled, in
keeping with the stereotype. His jacket provided him too little room, but at least his shoes were tied. His beard was full,
his hair curly. He had a meaty head and a broad nose. On
top of the nose sat a pair of tiny round spectacles. He clum-
sily managed to arrange his notes and demonstrations for the
day but eventually looked up, perhaps not too encouraged
by what he saw: a roomful of people who expected him to
speak. So he spoke. The voice, however, did not match the
figure, which was more oblate than prolate. One expected
this mountainous mess of a man to command a baritone's
register, but his voice was unnervingly high-pitched instead,
and his expressions were a bit affected.[1] He wondered if he
sounded nervous.

His audience was attentive, however, even as latecomers
scrambled for the few remaining seats. The class knew what
to expect and were willing to forgive the professor his awk-
wardness. He began by reviewing the previous day's findings,
a gentle enough start. But the blackboard was soon adorned
with its first equation, at which point everyone, including the
professor, settled in. The first equation was followed by many

others, with the most significant results recorded on a long blackboard that stretched across the front of the classroom; a few finer points were elaborated on two smaller boards on either side; and by the end of the hour, a sprawling mathematical edifice had been constructed.[2] He was methodical in his approach, attentive to every detail and to the needs of his students. He anticipated potential misunderstandings and welcomed their questions. He imparted meaning to the mathematics. These symbols refer to something real, he told them. He did make a small mistake at one point but caught the error quickly, saying, "Ach, that was stupid of me!"[3] The class chuckled. The tone was remarkably casual, even modern, especially given the circumstances, which were just barely twentieth century and decidedly European.

After finishing the lecture and bidding the class a good day, the professor lingered for a moment at the front of the room. Two students approached him, as was their habit. They asked him if he might settle a dispute. One of the students was a spirited young man named Paul Ehrenfest. He was clearly the sort of student who did not understand something until he had a chance to talk about it, and Boltzmann was happy enough to indulge him. Ehrenfest insisted that an unstated assumption lurked among the details of the day's derivation. The student who accompanied him, however, clearly disagreed, pointing to a few marks that she had made in the margins of her notebook. Her name was Lise Meitner. Boltzmann smiled, a grin buried beneath his beard. He would have benefited greatly had he known what these two students would one day accomplish. Meitner would eventually contribute to the discovery of nuclear fission, earning her a spot on the periodic table— the element meitnerium. And Ehrenfest's need to talk (particularly his gift for asking the right question) would place him within Einstein's inner circle. Boltzmann recognized the potential of these two students but never really appreciated his role in developing it. He told them that their debate could be easily settled, but joked that he was unwilling to do the settling, and asked that they perform an additional calculation. The physicists in training eagerly accepted his challenge and quickly headed in opposite directions. The professor sighed. He surveyed the now empty room, admired what he had managed to somehow create on those blackboards, and wandered out, content in knowing that he had survived yet another day.

16   One's Harshest Critic

Ludwig Boltzmann was a good teacher. If he were a twenty-first-century professor, he would have many favorable teaching evaluations on file. He was instead a professor of the late nineteenth and early twentieth centuries, a time when student evaluations of teacher performance were not yet the norm, so we have no hard data. We cannot say, for example, what percentage of turn-of-the-century Viennese physics students felt that they *learned something valuable from Professor Boltzmann's courses*. We have only anecdotal evidence of his professional competence. So there we turn. His assistant, Stefan Meyer, said, "Seldom was such outstanding teaching ability coupled with such extensive knowledge."[1] And Meyer was not alone in his assessment. Lise Meitner offered similar praise: "His lectures were the most beautiful and most stimulating that I have ever heard."[2]

But Boltzmann was recognized for more than just his intellect and communication skills. He also exhibited certain personal characteristics that endeared him to his students. One student described him as a charismatic figure: "One of my fondest memories from my youth are Boltzmann's lectures in the largest lecture hall, into which the students were jammed in a truly frightening way. This in spite of the fact

that I unfortunately grasped very little of his deductions. Like
me, countless came merely because of the radiant power of his
incomparable personality."[3] Another student was impressed
by Boltzmann's sweet nature. He wrote in a letter home to
Japan that "Professor Boltzmann is gentle and honest, and has
a personality to be loved by everybody rather in contrast with
his features."[4] Such feelings of goodwill were confirmed by
Boltzmann's colleagues, one of whom observed, "The unpre-
tentious great man made himself available to all for hours on
end, always equipped with unending patience and delightful
humor."[5]

If we were to translate this last observation into twenty-
first-century teacher evaluation parlance, we would say that
Boltzmann's students found him to be *accessible*, meaning that
he tended not to rush back to his office at the end of a lecture.
Or if he did rush back to the office, he at least resisted the urge
to close his office door. Which is to say that Boltzmann simply
did the right thing by his students. What is perhaps more sur-
prising is that many of Boltzmann's students found him to be
someone they could relate to (translated again into the twenty-
first-century vernacular). This sentiment was expressed by
Fritz Hasenöhrl, who was Boltzmann's student and eventual
successor at the University of Vienna. Hasenöhrl described
Boltzmann this way: "He never played up his superiority:
everyone was at liberty to ask questions and even to criticize
him. One could converse with him in an uninhibited way
as if between equals. ... He did not measure others with the
yardstick of his own greatness."[6]

This statement is remarkable in the sense that it is not at
all representative of its time. It is true that twenty-first-century
students often assume a tremendous level of familiarity with
their professors (even on the first day of school). But their

nineteenth-century counterparts did not enjoy such liberties. Higher education was decidedly undemocratic in nineteenth-century Europe. Interactions between faculty and students tended to respect the academic hierarchy, atop of which sat the likes of Herr Professor Boltzmann. Boltzmann's informality therefore was something of an anomaly.

Despite the fact that his disregard for decorum seems to have endeared him to his students, it was sometimes considered disruptive by other faculty members, especially outside his native Austria. In Berlin he found himself at odds with Helmholtz over the matter. Boltzmann described a visit to Berlin this way: "When I harmlessly adopted my usual tone on the first day in the Berlin laboratory, a single glance from Helmholtz made it clear that cheerfulness and humor did not befit the scholar,"[7] which seems to indicate that Boltzmann was aware of how his colleagues perceived him. It is not entirely clear, however, whether he was also able to accurately assess his students' perceptions. Did he know that his students considered him to be such a talented teacher? And did he really mean to treat them as equals? Why would he have given them that impression?

It's possible that Boltzmann was just a nice guy—the kind of person who did the right thing for the right thing's sake. But the situation may have been more complicated. Consider the circumstances. Boltzmann's classroom did produce a Meitner, but most of his students were in no way extraordinary. They were like the rest of us. We live; we die; we provide history with insufficient reason to be remembered. We might all be equals in the eyes of the law (yeah, right), but we are not all equals in the eyes of history. Boltzmann's intellectual contributions secured for him a place in history that eludes the 99.9999 ... percent of us. He was not a regular person, and he was likely aware of this fact. Then why would he have adopted such an egalitarian spirit in the classroom?

To answer this question, we turn again to our postmortem diagnosis. We reaffirm the view that Boltzmann suffered from an anxiety disorder. But anxiety disorders appear in multiple forms, and I've not yet attempted to give Boltzmann a more specific diagnosis. We're guided, however, by clues from his behavior both in and around the classroom. Boltzmann clearly cared what his students thought of him and frequently worried that they might judge him unfavorably. Much of his anxiety, in fact, seems to have focused on his classroom performance. His wife, Henriette, documented this struggle, writing to their daughter Ida that "Papa is neurasthenic the nights before lectures,"[8] which would have been a problem for someone whose primary professional responsibility was to deliver lectures. A colleague in Leipzig, Wilhelm Ostwald, reported similarly that "despite the grateful reception by his new students he was beset in Leipzig by the most serious ailment that a professor can encounter: fear of lecturing."[9] And we know that Ostwald's assessment was not mere conjecture because Boltzmann had written to him before arriving in Leipzig, saying, "I have never concealed that I have a tendency toward nervousness."[10] His anxiety was pervasive enough that he even mentioned it in a memoir. "Now I must confess," he wrote, "that I always suffer a little from stage-fright before the first lecture,"[11] which of course makes very little sense. Boltzmann was a celebrated scientist who had no reason to fear the lecture hall. But unfortunately, reason is not always enough. Sometimes we struggle when we should thrive. And sometimes our self-esteem hinges tenuously on a feeling that we might be held in high esteem.

The problem described here is a social anxiety, a social phobia that seems to have followed Boltzmann around and may have even chased him to his grave. But before we trace his path to self-destruction, let us pause and paint this cloud a silver lining. Boltzmann's psychological burden, albeit a personal hardship, likely benefited his students. Why? Or maybe this is a *how* question. We turn in any case to Lise Meitner, who displayed an almost unreasonable degree of insight on the matter. She said, "He may have been wounded by many things a more robust person would hardly have noticed. ... I believe he was such a powerful teacher just because of his uncommon humanity."[12]

Meitner seems to suggest here that Boltzmann was *too* sensitive, that he cared too much what other people thought of him. And her statement reveals how an apparent handicap can also be an advantage. It's often good to care what other people think of us. It demonstrates a respect for others and for our relationships with them. It motivates us to satisfy their expectations and to exceed those expectations whenever possible. Good teachers inevitably share this concern. But good teachers generally do not suffer from it. The tendency to care what other people think becomes an impediment only when it blooms in excess or when it imposes on us a consistent need for approval, so that when we find ourselves in front of a classroom (or boardroom, or courtroom, etc.), we grant the situation more weight than it deserves. But this tendency need not be a cause for shame. As Meitner pointed out, these motivations and concerns are characteristically *human* tendencies. Now, admittedly, they might not be uniquely human tendencies. Chimpanzees certainly care about their standing among other chimpanzees, and some chimpanzees

probably care too much about such things. But human interactions are so complex and human self-awareness is so highly
(almost unreasonably) developed that social anxiety seems
almost to be part of our human condition. And that condition is routinely threatened by the fact that human self-worth
is inescapably tied to how we are perceived by others, which
is something that we can never really know. That last part
can be truly maddening. What do other people think of us?
How important is the answer? It often seems that we have no
choice but to guess, and we often guess incorrectly. Such a
dilemma was not Boltzmann's alone. It is a human dilemma—
one that we negotiate as best we can, knowing that many of
life's pitfalls and rewards hang on it. And as Boltzmann tragi-
cally demonstrated, we cannot easily opt out of this dilemma
without actually opting out.

17 Something Like a Mathematical Supplement

I admit with some reluctance that you might not need to read this chapter. I say *with some reluctance* because the comment probably seems a bit odd. Why did I bother to include a chapter that might be skipped? Let me explain. The challenge is that different readers will know different things. Some people who read a book about entropy will inevitably know a lot of mathematics, admittedly much more than what they'll ever encounter in this book. But other readers will require more mathematical background. If, for example, you're many years removed from any mention of a logarithm, then you might find the next few pages to be useful. But if you already know about logarithms and other such things, then skim or skip or use your own best judgment.

So now that everyone has chosen his or her own adventure, let us proceed with an example. Consider the following expression: $\log_{10}(100)$. This expression tells us to take the logarithm of 100 to base 10, which prompts us to answer the question: *10 to what power gives 100?* Our task then is to solve the equation $10^x = 100$ for x. Or perhaps more simply put, *How many 10s need to be multiplied together in order to get 100?* Now reasonable people can disagree about many things but not

Table 17.1
Examples of the logarithm with base 10

Question	Answer	Related statement
$\log_{10}(1) =$	0	$10^0 = 1$
$\log_{10}(10) =$	1	$10^1 = 10$
$\log_{10}(100) =$	2	$10^2 = 100$
$\log_{10}(1000) =$	3	$10^3 = 1000$

about the answer to this problem. We need two 10s or $x = 2$ because $10^2 = 100$ and so $\log_{10}(100) = 2$. And we can make any number of such statements. Table 17.1 has just a few examples.

The only one of these examples that might require some further explanation is the case $\log_{10}(1) = 0$, which implies that $10^0 = 1$. You can, in fact, take any number (other than 0) and raise it to the 0 power and get 1, which suggests that you don't need to specify the base for the case $\log(1) = 0$. Here we'll work with base 10 because it's relatively common (the logarithm to base 10 is actually called the common logarithm) and because it provides a potentially useful measure of the number of digits in a number. In the examples in the table, the logarithm of the number is simply one less than the number of digits in the number. (There are three digits in the number 100, and its logarithm is 2.) The examples here are also especially straightforward in that the logarithm of the number is equal to the number of 0s in the number. (There are two 0s in the number 100, and its logarithm is 2.)

The overall effect, therefore, is that the log function has the ability to rein in runaway big numbers. The number 1,000,000 may appear significantly bigger than the number 10, but the logarithms of those two numbers are not so different. The logarithm of 1,000,000 is 6, and the logarithm of 10 is a comparable value of 1, which could be a useful property in a book with some really big numbers. Those numbers might appear to be somewhat less extraordinary after being subjected to the transformative powers of the log function. (Actually, some of the numbers in this book are still pretty extraordinary even after being fed through the log function.)

Yet why is the log function any more useful than any other topic from a long-forgotten math course? The problem here is that math courses, like many other valuable experiences, are generally wasted on the young, and young people often don't know what it is that they might one day need to know. This situation is particularly troubling given the fact that we're all guilty at some point in our lives of being a young person. If only our future selves were able to review our syllabi beforehand and recommend to us days when it might be particularly important to pay attention in class. If that were possible, we would learn that some seemingly inconsequential topics will one day prove to be useful. The log function, for example, is necessary for our purposes here because it satisfies the following identity (provided that both a and b are positive):

$$\log_{10}(a \times b) = \log_{10}(a) + \log_{10}(b)$$

We can check that this identity holds by substituting numbers for the variables. Substituting the numbers 10 and 100 for the variables a and b, respectively, we have:

$$\log_{10}(10 \times 100) = \log_{10}(10) + \log_{10}(100)$$
$$\log_{10}(1000) = \log_{10}(10) + \log_{10}(100)$$
$$3 = 1 + 2$$
$$3 = 3$$

The mathematics makes the case quite succinctly, but we can also dress the mathematics in words. Two numbers, called factors, can be multiplied together to give a new number, called a product. The product on the left is 1000, and the logarithm of the product is 3. The two factors on the right are 10 and 100, whose logarithms are 1 and 2, respectively, which, when added together, also give a value of 3. We can state this result more generally: the logarithm of a product is equal to the sum of the logarithms of the individual factors. This particular property of the log function, its ability to turn products into sums, may appear insignificant, but it's the reason that S equals $k \log W$ and not something else. The physics follows the mathematics. And we now have a direction in which to head:

$$\log(W_A \times W_B) = \log(W_A) + \log(W_B)$$

18   You May Now Return to Your Seats

There's no need for a spoiler alert. We already know that $S = k$ log W, and we know that Boltzmann killed himself. We've known these things from the start, although we don't yet know all of the reasons. Why exactly does S equal k log W? And why did Boltzmann kill himself? Was his life really that bad? There must have been more to the story.

Let us first return the question of the log function. The entropy S is a function of the multiplicity W. According to the equation, S increases as W increases. The entropy therefore is expected to be high when the system occupies a macrostate of high multiplicity. But there are lots (in fact, an infinite number) of functions that exhibit a similar trend (the trend in which S goes up as W goes up). Why not use an even simpler function like $S = kW$? Boltzmann had good reason to include a logarithm in his equation, and we can now discuss the reason.

Previously in the book, we considered a room in which the atoms were located either on the left- or the right-hand side and used the positions of those atoms to calculate the multiplicity of the available macrostates. But we can do better than to contemplate the fate of a single room. We can expand our horizons by expanding the size of our system. Let us therefore consider not one but two rooms. (We've decided to expand our horizons incrementally.) Our goal is to calculate the total entropy of the two rooms.

Now we don't want to overthink this problem. Our instincts are likely to be instructive here, so we might first consider a few related problems for which we already know the answers:

Question 1: *How would we calculate the total number of atoms in the two rooms?*

Answer: We would add the number of atoms in one room to the number of atoms in the other room.

Question 2: *How would we calculate the total volume?*

Answer: We would calculate it the same way—by adding together the two individual volumes.

Question 3: *What about the total energy?*

Answer: Energy is a bit more abstract a concept, but it's similar in this respect. The total energy is just the sum of the two rooms' energies.

In each of these cases, we simply need to add, and entropy is no different. We simply need to add together the two entropies. The number of particles, volume, energy, entropy—these properties are all additive. And these properties are additive not just for two rooms but for any number of rooms, so that if we were to add a third, fourth, and fifth room to the system, we would simply add a third, fourth, and fifth entropy to the total entropy.

But what about the multiplicity? We need to determine the total multiplicity of the two rooms and relate that quantity to the total entropy. A few additional details might help to make this problem more concrete. Assume that one of the

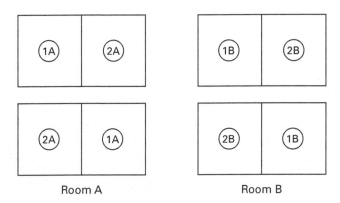

Figure 18.1
Microstates for the two individual rooms

rooms, called room A, has two atoms, 1A and 2A, while the other room, called room B, has two atoms, 1B and 2B. If each room occupies its macrostate of highest multiplicity, then each room has one atom on the left and another atom on the right. The multiplicity of each *individual* room therefore is 2 (figure 18.1).

The *total* multiplicity, however, is obtained by combining each of the microstates for room A with each of the microstates for room B (figure 18.2).

Given the fact that each room has two microstates (in its macrostate of highest multiplicity), the total multiplicity is just $2 \times 2 = 4$. The multiplicities multiply. The outcome is relatively straightforward in this case because room A's microstate does not influence that of room B. This assumption—that each room realizes its microstate independently—allows for the combination of microstates through simple multiplication.

Here then is the scenario that we face. The entropies of the two rooms add, while the multiplicities multiply. How can we

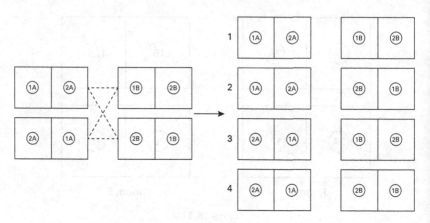

Figure 18.2
Combining the microstates in order to determine
the total multiplicity

satisfy both of these criteria? By invoking the log function.
Call the multiplicities of the individual rooms W_A and W_B
and the total multiplicity W_T. We know that

$$W_T = W_A \times W_B$$

And if W_T is equal to $W_A \times W_B$, then we know that $\log(W_T)$ is
equal to $\log(W_A \times W_B)$ (because we can take the log of the left-
hand side of an equation so long as we take the log of the
right-hand side):

$$\log(W_T) = \log(W_A \times W_B)$$

Once we introduce the log function, we can then call on
its ability to turn products into sums:

$$\log(W_T) = \log(W_A) + \log(W_B)$$

Then the only remaining step is to multiply through by
the missing constant k:

$$k\log(W_T) = k\log(W_A) + k\log(W_B)$$

We now have an expression on the left that looks like the total entropy and two expressions on the right that look like the entropies of the individual rooms. The total entropy therefore is just the sum of the individual entropies:

$$S_T = S_A + S_B$$

We started with the condition that the multiplicities multiply, and we arrived at a result in which the corresponding entropies simply add.

This result required a few equations, which is not always a popular move in a popular science book. Science icon Stephen Hawking has typically avoided equations in the books that he's written for a general audience. He claims that each equation cuts a book's sales in half. His classic *A Brief History of Time* included just one equation,[1] and that equation was $E = mc^2$ (which maybe should have been considered a gimme). If we are to believe Professor Hawking, who is, after all, a best-selling author, this particular book is now doomed to failure. But the good news is that the math has done its work, and we can now reap its many benefits.

19 Boltzmann's Constant

Boltzmann's 1877 paper—the publication whose findings we celebrate here—is long.[1] It's 63 pages, and its title alone is 166 characters (including spaces and not discounting for the fact that it includes some words that sound ridiculous to all but the most German of ears). If Boltzmann were to submit this paper for publication today, the editor would kindly remind him of the journal's page count restriction and suggest that some of the content be moved to the journal's online supplementary materials. And yet despite its length and the significance of its main result, you'll find nowhere in it the equation $S = k \log W$. One might expect the equation to emerge triumphantly at the end of the paper, accompanied by some memorable pronouncement for the ages. But the idea, if not the actual equation, is found by only the most diligent readers (who likely also have some graduate-level training in either chemistry or physics).

Although the essential idea is rightly attributed to Boltzmann, the equation was not written in so compact and lasting a form until Max Planck's use of it in 1901.[2] Planck was fourteen years Boltzmann's junior and interested not so much in the study of gases as he was the study of light emitted by hot objects. He found, however, that entropy was central to both subjects and described his investigations this way: "This quest automatically led me to study the interrelation of entropy and probability—in other words, to pursue the line of thought inaugurated by Boltzmann. Since the entropy S is an additive magnitude but the probability W is a multiplicative one, I simply postulated that $S = k \log W$, where k is a universal constant."[3]

This brief description of the reasoning that led Planck to $S = k \log W$ represents an amazing distillation of Boltzmann's much longer paper. Boltzmann was either unable or simply just not inclined to articulate his own ideas so clearly. One might even ask, if Boltzmann were to rise from the grave, how he would regard the equation that adorns his monument in the Central Cemetery. He would certainly recognize the idea as his own. But would he claim sole ownership for its inception? Or might he be willing to acknowledge that the equation bears another man's stamp?

It might not really matter (except to those of us for whom it matters). The textbooks have been written and the credit assigned to Boltzmann, and there appears to be little (and by "little," I mean "no") public outcry to install some footnote or citation alongside Boltzmann's grave. Nor is there any reason to feel that Max Planck was somehow aggrieved in the matter. Planck was awarded a Nobel Prize for what he was able to accomplish with $S = k \log W$, while Boltzmann was only nominated (and by Planck of all people).

It was a complicated relationship, both intellectually and personally. For much of his career, Planck was actually an outspoken critic of Boltzmann. Planck's scientific views were rooted not so much in atoms and statistics but in the tangible objects of classical thermodynamics—things like steam engines and ovens that were governed by firm, immutable predictions. Boltzmann's entropy must have seemed a flimsy notion in comparison. His methods could predict nothing more than the likelihood of an event, an unsettling proposition for a man and a physics community not yet accustomed to uncertainty. How then did the conservative Planck manage to adopt Boltzmann's methods? The answer, it seems, is that Planck simply adhered to the scientific method, which requires its followers (of which there are too few) to give up deeply held beliefs when the evidence demands it. Planck was wise enough to respect this doctrine of the scientific faith, although he was a reluctant (almost penitent) convert to Boltzmannism, turning to statistics only after all other methods had failed him. Writing to a friend, he described his conversion this way: "Briefly summarized, what I did can be described as simply an act of desperation. ... I was ready to sacrifice every one of my previous convictions about physical laws."[4]

After he emerged, however, blinded and stumbling from his personal road to Damascus, Planck was eager to receive the elder Boltzmann's approval. He even acknowledged Boltzmann's blessing during his Nobel lecture. In that lecture, he referred to a correspondence with Boltzmann from around the time of his discovery. "It brought me much-valued satisfaction," he said, "for the many disappointments when Ludwig Boltzmann, in the letter returning my essay, expressed his interest and basic agreement with the train of thoughts expounded in it."[5] Planck reported that Boltzmann's attitude toward him shifted thereafter from "ill-tempered" to "friendlier,"[6] which must have seemed a promising enough baby step in the right direction. But Boltzmann hung himself by a noose a few years after this easing of tensions, and so their personal relationship would never reflect what their intellectual alliance had accomplished.

Several years later, Planck still seemed ambivalent about how to regard Boltzmann (perhaps because he knew that Boltzmann never regarded him very highly). In his autobiography, he wrote, "I could play only the part of a second to Boltzmann—a second whose services were evidently not appreciated, not even noticed by him,"[7] a seemingly odd statement from a Nobel laureate, especially one who was recognized in his lifetime as a founding father of quantum mechanics—a field that would forever change our view of the world. Here, Planck sounds more like an emotionally neglected child than a scientific authority figure.

There was, however, one matter in which Planck never hesitated to assert his priority over Boltzmann: Planck resolutely claimed credit (and rightfully so) for the constant k that precedes the log W. Referring to k in his Nobel lecture, he wrote, "This constant is often referred to as Boltzmann's constant, although, to my knowledge, Boltzmann himself never introduced it—a peculiar state of affairs, which can be explained by the fact that Boltzmann, as appears from his occasional utterances, never gave thought to the possibility of carrying out an exact measurement of the constant."[8]

Now it seems unfair to assume that Boltzmann never gave any thought to the matter. He certainly would have recognized the need for a constant in practical matters, in applications of a particular sort. But Boltzmann was not always engaged in practical matters, and he often left doors wide open through which other scientists then passed. Planck, in contrast, tended to tie his theories more closely to experiment, and he was also sensible enough to recognize when a detail was an especially important one that would merit a permanent place in the scientific canon. Yet despite his many insights, the constant k does not bear Planck's name. That honor was given to Boltzmann instead, the man whose occasional utterances supposedly never gave any thought to the matter.

We should not, however, be too quick to assign Planck the role of the victim here. It is true that some scientists see further than others, but they do so by standing on the shoulders of giants. Planck himself was a giant, so sometimes we have giants standing on the shoulders of giants, which can lead to several competing claims for priority. Do we, in this case, favor Boltzmann, whose shoulders were especially broad? Or Planck, who somehow managed to climb on those shoulders? The situation is further complicated (but also resolved) by the fact that Planck's discovery introduced not one, but two, fundamental constants. His derivations started with $S = k \log W$, but they ended with $E = h\nu$, an equally historic finding whose constant h would also need a name attached to it. If, in retrospect, we assume (quite rationally) that Planck's name could not (without considerable confusion) be associated with two fundamental constants, then it seems a fair result that Planck was given h and k was given to his predecessor. Neither man, therefore, was aggrieved, and both names secured a place on the inside cover of every physical chemistry textbook, along with the values $k = 1.381 \times 10^{-23}$ J/K and $h = 6.626 \times 10^{-34}$ J·s. We'll leave Planck's constant h for some other book, except to note that it, like k, is a very small number. Very small numbers are needed in order to account for very small things, and atoms are very small things.

Let us focus then on k, which tells us something about the kinetic energy of those very small things. It does so by conferring on entropy its units, which are commonly reported as joules per kelvin (J/K). The joules refer to energy, while kelvin (abbreviated with a capital "K") refers to temperature, so we have units of energy divided by units of temperature. The fact that Boltzmann's constant is charged with the task of relating energy with temperature is actually something of a historical artifact. Temperature was originally defined in operational terms. Thin glass tubes were filled with mercury or some other substance whose length inside the tube varied in some regular way with the perceived hotness or coldness of the surroundings. (That sounds like a much too complicated description of a thermometer.) Low temperatures were assigned to cases in which the column of mercury was short and high temperatures to cases in which the column was long.

In order to facilitate comparisons between measurements, standard temperature scales, such as those of Celsius and Fahrenheit, were eventually adopted. The Celsius scale used the freezing and boiling points of water to calibrate its 0 and 100 degree marks, while the Fahrenheit scale relied on the freezing point of saltwater and the average human body temperature for its 0 and 100 degree values. A more theoretically motivated system was later developed by Lord Kelvin, whose scale was tied not to water or saltwater or anything so anthropocentric as the human body. His scale was general, and it was absolute. Its lower limit really was 0. There is nothing absolute about 0 °C or 0 °F, but 0 K is a true limiting value, and for that reason, most physical scientists (meteorologists notwithstanding) prefer the Kelvin scale. It is important to note, however, that the Kelvin scale was not a complete reworking of all that preceded it. It actually shares a common size increment with the Celsius scale, which means that a change of 1 kelvin is equivalent to a change of 1 degree Celsius—that is, if the temperature increases (or decreases) by 1 kelvin, then it increases (or decreases) by 1 degree Celsius. The Kelvin scale is simply shifted relative to the Celsius scale in order to establish a more meaningful 0 point (figure 19.1).

From these relatively simple measurements, the field of thermodynamics grew, and a theoretical basis for temperature gradually emerged, eventually culminating with the discovery that temperature is proportional to the average kinetic energy of the atoms in a gas. Temperature therefore is a measure of kinetic energy. It's high when atoms have lots of kinetic energy and low when the atoms have relatively little. It really

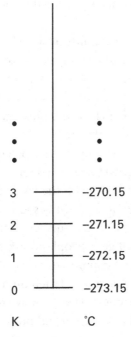

Figure 19.1
Relationship between the Kelvin and Celsius scales

doesn't need its own set of units. Rather than some number of kelvins (or degrees Celsius or degrees Fahrenheit), what we really need are some number of joules (or calories or electron volts). Units of energy would suffice and would more faithfully represent the true nature of things. And they could, if they were to replace our standard units of temperature, tie our everyday experiences more closely to the microscopic goings on around us. Why is a warm summer day warm and a cold winter night cold? Because there are atoms all around us, and the motions of those atoms reflect the energy of the season. That last idea is probably a little bit hokey, but it's worth noting that our daily weather forecasts mislead us in thinking that nothing has been learned since the time of Daniel Fahrenheit (who died in 1736).

Hindsight, however, has little chance here of trumping history. If temperature is the thing that's most readily observed in the study of thermodynamics, then maybe it's only fitting that the quantity carries with it some historical baggage. For arguably, no other branch of the physical sciences has such a convoluted history as the study of thermodynamics. P. W. Bridgman was a great champion and practitioner of the subject, who also happened to be the recipient of the 1946 Nobel Prize in Physics. For Bridgman, the laws of thermodynamics provided the basis for a remarkably successful career, and yet he was keenly aware that the subject often prompted many thoughtful and intelligent people to ask themselves if they really understood things properly. He wrote that the laws of thermodynamics "smell more of their human origin" than do the other laws of physics.[9] These laws were not bestowed in some perfect form on some faithful prophet who sat unwittingly on some fateful mountaintop. They were unearthed gradually by a dedicated community of scientists whose

collective insights rarely traced the shortest path from ignorance to enlightenment. The result was not so much a set of
platonic ideals that had awaited our discovery as it was a set of
empirical rules that matched our observations. And those rules,
Bridgman noted, are "more palpably verbal" than are the other
laws of physics,[10] which has also alienated the subject from its
target audience. Why should a group whose native language
is mathematics assume ownership of a subject that relies so
heavily on words in order to make its arguments? Words are
blunt instruments that can provide cover for shoddy reasoning. Surely some contradiction must lurk somewhere among
all of that verbiage.

This little book aspires to explain just one critical part of
the whole thing: the entropy as Boltzmann came to understand it. But we share Bridgman's assessment of the broader
subject and admit that, just as Boltzmann was not without
his flaws, so too the subject of thermodynamics is not without its idiosyncrasies. I recognize that the Kelvin system, as well
as those of Celsius and (god forbid) Fahrenheit, will continue
to enjoy a historically protected status, but I also contend that
these traditions have imposed on us a somewhat unwelcome
inheritance. If our units of temperature are not to reflect the
underlying kinetic energy of the atoms, then some other quantity needs to reestablish that connection. That some other
quantity is Boltzmann's constant.

Although entropy was introduced long after temperature,
its arrival coincided with and even contributed to the recognition that temperature is a measure of kinetic energy. The form
that it took therefore was dictated by the needs of the system
into which it was incorporated. Let us review. We started with
W, the number of microstates associated with a particular
macrostate. The important thing to note, for our purposes, is

$$\frac{\text{units of energy}}{\text{units of temperature}} \times \text{units of temperature} = \text{units of energy}$$

$$\frac{J}{K} \times K = J$$

Figure 19.2
Boltzmann's constant converts units of temperature
to units of energy

that W is just a number. When we then took the log of W, the result was still just a number, but now when we multiply by k, the result is no longer simply a number; it's a number attached to a particular set of units: units of energy divided by units of temperature. Consider then what happens when entropy or simply k is multiplied by temperature (figure 19.2).

All that remains are units of energy. This operation returns temperature to its more natural state, converting it from some number of kelvins to some number of joules. The conversion factor in this case is Boltzmann's constant, and the operation is a routine procedure. Scan any physical chemistry textbook and you'll find scattered throughout its pages: $pV = NkT$ and $G = H - TS$ and $\exp(-E/kT)$ and lots and lots of other examples in which a T is closely accompanied by either a k or an S. But remember that we would not need this operation if not for the fact that the true identity of temperature is obscured by the symbols K, °C, and °F that stubbornly follow it around. Boltzmann's constant would be unnecessary, and entropy would be simply log W if not for our customs, which are subject to the whims of our history.

But maybe we (*we*, in this case, meaning all past, present, and future members of the species *Homo sapiens*) should be proud just to have made it this far: to have devised a system that, if not admired for its clarity, actually works and to have created a set of tools, both real and conceptual, that meet the intellectual demands of a world that's increasingly consumed with the consumption of energy. And if Boltzmann's constant appears, in retrospect, to be something that maybe we could live without, why at this point would we want to deprive ourselves of it? Might we not derive some imperceptible benefit from it? After all, it reminds us that a man named Boltzmann, someone we might not otherwise have any reason to remember, once thought about matters that still matter to us today. And it connects us to that someone by inviting him to join us (if in name and spirit only) in the calculations that he inspired.

Precipitating events led to Boltzmann's decline: the fumbled job negotiations, an unsuccessful appointment as university rector (a role similar to a modern university president), and the death of his mother and his first-born son, who was also named Ludwig. Self-doubt was not immediately incapacitating, however. It wore slowly on his psyche, until one day he was overwhelmed by routine tasks and struggled even to enter the classroom. Over the course of that long slide, Boltzmann contended with foes who were much more formidable than his freshman physics students. He wrangled with respected colleagues who looked suspiciously upon his methods.

The tools of statistics and probability were neither widely adopted in the nineteenth century nor were they trusted. They were part of a then suspect branch of mathematics whose only known application was in gambling, obviously a disreputable activity. (Today it is not so much disreputable as it is profitable—for casino owners and local governments.) The Scottish physicist James Clerk Maxwell was, like Boltzmann, a pioneer in the use of statistics who, unlike Boltzmann, recognized the bias against it. In a letter to a friend, Maxwell wrote: "The true Logic for this world is the

Calculus of Probabilities. ... This branch of Math., which is generally thought to favour gambling, dicing, and wagering, and therefore highly immoral, is the only 'Mathematics for Practical Men,' as we ought to be."[1]

But how exactly does an affinity for statistics distinguish a practical man from any other? A practical man, Maxwell seemed to argue (we presume that he would have argued similarly on behalf of practical women), adopts the tools that are needed in order to answer his question. He finds something that works. And when those tools stray from the ordinary, the practical man anticipates the opposition that will inevitably follow because the practical man knows that science is not some disembodied search for the truth. It proceeds on the backs of scientists, whose motivations are many. It seeks something that's increasingly unassailable in the long term but bends toward the biases of the prevailing authorities in the short term.

For Maxwell, those biases were the terms and conditions to which he agreed as a practicing scientist. But for Boltzmann, these human factors were the source of considerable frustration. He would have much preferred to participate in a disembodied search for the truth (as long as he received credit for the truths that he himself uncovered, of course). In this respect, Boltzmann's attitude was frequently described as naive and his manner childlike. His friend and colleague Wilhelm Ostwald called him "a stranger in this world. Constantly occupied by the problems of science, he had no time or inclination to deal with the thousand trivialities that take a large part of the life of modern man, and which he handles instinctively. This man, whose mathematical acuity would not pass over the slightest inconsistency, showed in daily life the innocence and inexperience of a child."[2] A family friend

offered a similar assessment, describing Boltzmann as "the pro-
totypical unworldly scholar, living wholly in the realm of his
science and his groundbreaking research. ... He commanded a
broad range of general knowledge, which had no impact whatever on the manifestly childish naïveté of his nature, as one
often finds with those whose focused minds move in higher
spheres."[3]

Yet despite his shortcomings, many seemed willing to tol-
erate Boltzmann, and some may have even appreciated his
adherence to stereotype. He was, admitted a fellow faculty
member, simply a "shy eccentric."[4] And he was often amusing,
although not always intentionally so. Consider, for example,
the time that he was seen wandering the streets of Graz alongside a cow that he had purchased at a local market.[5] Boltzmann
had apparently decided that his children needed fresh milk,
so he bought them a cow, not knowing exactly how to milk it
or what to feed it or even how to get it home (assuming that
he knew his way home). He later consulted a zoology professor (without realizing that he might have had better luck had
he befriended a local farmer instead).

The fact that this charming portrait of a man (and his cow)
is recounted still today suggests that many of Boltzmann's
contemporaries considered his eccentricity to be a source of
mild amusement and not a serious character flaw. But the
record also clearly indicates that a few found him to be less
than lovable. One colleague described him as a "powerful
man, but childlike to the point of childishness."[6] And another
claimed, "Boltzmann is not malicious, but incredibly naïve
and casual ... he simply does not know where to draw the
line."[7] This last comment appeared in a letter written by Ernst
Mach, the physicist-turned-philosopher who disapproved not
only of Boltzmann's disposition but of his entire life's work.

Mach's critique of Boltzmann involved very little of what we would currently consider to be physics. It was certainly not a debate over technical matters, as is often the case today. Modern scientists tend to challenge one another mostly over issues of protocol and interpretation, asking questions of the sort: *Is a particular approximation justified?* or *When does a signal emerge above the noise?* They tend not to engage one another over philosophical matters. You can ask them, if you'd like, to describe for you their philosophy of science, but don't be disappointed if you're not subsequently engaged in an engaging conversation. You're more likely to be greeted with a squirm or a snort or a sneer.

But many nineteenth-century scientists were not yet ready to relinquish the title of natural philosopher, preferring instead to retain ownership of those lofty (and generally intractable) questions: *What is this thing that we study? What is a valid method of inquiry? How can I reconcile my observations with those of another?* These were the questions that Mach hurled at Boltzmann and that Boltzmann picked up and examined— hoping to defuse them. But Boltzmann was ill suited for phi- losophy (and for philosophers). His talents and temperament were those of a modern physicist.

The central issue in the case of Mach versus Boltzmann was the atomic theory, which posits that matter is composed of particles that are too small to be seen. And it is that last part—the *too small to be seen*—that provoked the controversy. For Boltzmann, whose eyes were much weaker than his intel- lect, it must have seemed self-evident that Mother Nature would prove to be more subtle than his senses could discern. Biological bandwidth is limited, after all, even among our most keenly evolved species. And why would he or anyone else have assumed that physics ends where our senses fail? Physics is not constrained by biological boundaries, nor are we. In fact, we have a long tradition of reaching beyond those boundaries and building instruments that render unobservable phenomena observable. Yet despite our ingenuity, some phenomena will always lie just out of reach. How should we account for them or for those things that are not directly observed but only strongly suggested?

Mach's response was simple: we do not account for such things, and we should not. His approach was disciplined, and it was austere. Establish a set of rules that explain observable phenomena and stop there. We know, for example, that the volume of a gas doubles if its temperature doubles (holding the other relevant variables constant). We know this trend to be true because many people have performed the experiment many times. We do not know the reason, Mach would argue, because the reason is unobservable, so we should not expect to know it. And the trend holds regardless of whether we have any deeper understanding of the mechanism.

It was an intellectually impoverished view of science (insert editorial disclaimer) that denied scientists the opportunity to answer the one question that motivates them most: *Why?* But it was philosophically pure and almost beyond reproach. Boltzmann could point to the gains that were made by his theories, but Mach could always argue that those gains were based on unprovable assumptions. What good is progress, Mach would argue, if it relies on fantasy? And how much do we really understand if we accept unknowable premises? It was not an unpopular view, especially among German-speaking scientists. It was rigorous and demanding and therefore admirable, and it afforded its adherents the moral high ground.

Let Boltzmann, the wayward atomist, succumb to his convenient imaginings. Mach and his disciples would uphold something sound instead.

Many younger scientists considered this view especially attractive—youth being more susceptible to philosophical extremism. Even Professors Planck and Einstein fell under Machian influences during their formative years, although both discarded those influences soon after trying them on. They wisely rejected Mach's call for the self-imposed sterilization of the scientific method. If they had not, their names would not be known to us today. Planck's legacy might have been that of a humble schoolteacher, and Einstein would likely never have made it out of the patent office. But they're remembered instead as founding fathers of modern physics, and they have Boltzmann (among others) to thank for the positions to which they were able to rise. Boltzmann fought the fight that they would be spared, securing permission for them to reference objects and forces that are visible to the mind's eye only.

If in retrospect this outcome seems to have been inevitable, it is only because our physics is so overwhelmingly aligned with that of Boltzmann. Our physics is Boltzmann's physics, while Mach's concerns have been relegated to philosophy departments (and other places where physicists rarely wander). The debate that surrounded these two men, however, was likely necessary, even if it now appears to be long forgotten. History needed a Mach, and it needed a Boltzmann. Someone needed to challenge the atomic theory, and someone needed to defend it. Only then could we have known the ground rules—what assumptions would be permitted and what the expected return on investment would be for those assumptions.

Someone also needed to simply hold the line while the evidence slowly mounted in favor of the atomic theory. A unified consensus was eventually reached, but its development was incremental, with some segments of the scientific community subsumed into the fold much earlier than others. Those scientists we today would recognize as chemists were (perhaps not surprisingly) the early adopters. They embraced the atomic theory as their guiding principle and fashioned from it a major branch of modern science. The British scientist John Dalton led their charge with his law of multiple proportions. He found that the elements combine with one another subject to certain constraints. The element carbon, for example, can combine with the element oxygen, but only in amounts corresponding to certain whole number ratios. If one starts with 100 grams of carbon, it can combine by way of one reaction with 133 grams of oxygen or by way of another reaction with 266 grams of oxygen—the significance of which is that the ratio of the two oxygen masses (266:133) reduces to 2:1. These reactions are related to one another by way of a small whole number ratio, which implies that the oxygen is packaged somehow into discrete amounts. A sensible explanation for this result is that matter is not forever divisible, but it is composed of some fundamental set of units or atoms. An atom of carbon can combine with one atom of oxygen to form carbon monoxide (CO) or with two atoms of oxygen to form carbon dioxide (CO_2), hence the 2:1 ratio. Later experiments were needed in order to determine how many carbon atoms are actually present in a 100 gram sample, but John Dalton's work suggested that 133 grams of oxygen provide one oxygen atom for every carbon atom in the sample (thereby making carbon monoxide), while 266 grams of oxygen provide two oxygen atoms (making carbon dioxide).

This law of multiple proportions appeared in volume 1 of Dalton's book *A New System of Chemical Philosophy*, which was published in 1808 and served as the basis for discoveries by Amedeo Avogadro and Josef Loschmidt and many other scientists who actually preceded Boltzmann. In fact, by the time that Boltzmann discovered $S = k \log W$, estimates were already available for both the size and the number of atoms that are present in a typical gas. His discovery therefore was not so much a part of Atomic Theory 1.0 as it was a feature that was added to the system's first major update. One could even argue that Boltzmann was not so much a revolutionary as he was a stalwart of an emerging consensus view. He was certainly not a lone revolutionary, nor was he the sole target of Mach's critique. He had many natural allies but failed to rally with those allies and imagined himself instead to be an army of one.

Among the many dangers of being an army of one is a potential inability to recognize when the war has actually ended. An individual can continue to fight much longer than necessary and may even confuse victory with defeat. Boltzmann, for example, did not seem to appreciate the fact that his study of gases was widely regarded as a significant achievement. He knew it only to be controversial and therefore considered it to be a failure. In part II of his *Lectures on Gas Theory*, he wrote, "In my opinion it would be a great tragedy for science if the theory of gases were temporarily thrown into oblivion because of a momentarily hostile attitude toward it. ... I am conscious of being only an individual struggling weakly against the stream of time. But it still remains in my power to contribute in such a way that, when the theory of gases is again revived, not too much will have to be rediscovered."[8]

Here we find Boltzmann indulging in all-out, self-pitying melodrama: "a great tragedy," "thrown into oblivion," "only an individual," "struggling weakly against the stream of time." He had apparently forgotten the many accolades that he had accumulated over the course of his career and felt compelled instead to insert his own personal struggles into his scientific writing. If he were alive today, he would be that aggrieved person who clicks Send late at night rather than first sleeping on the matter.

Boltzmann also seemed to have been unaware of the fact that big ideas merit lots of attention and that lots of attention often leads to scrutiny. The challenge, of course, is to survive that scrutiny with one's ego intact, the likelihood of which varies considerably from person to person. There are some people who think highly of themselves regardless of the circumstances and for whom criticism might not to even register. (These people are easy to find. Just check C-SPAN.) At the other end of the self-awareness spectrum, say a little bit further down the road to enlightenment, are people who consider scrutiny to be painful but manageable and might even value the role of constructive criticism. (These people are generally much harder to find, and they almost never seek elected office.) Then there are the Boltzmanns of the world—people whose waters are less still or run less deep (or both) and who consider criticism of any kind to be painful. These fragile types are easily wounded by threats both real and imaginary and are sometimes too sensitive even for places like the classroom. They tend to see debate not as an exchange of ideas, but rather as a series of personal attacks, and they often struggle to evaluate situations objectively. Which then prompts us to ask: Should we feel any sympathy for Boltzmann? Or should we blame him instead for his constitutional shortcomings? Or might the truth lie somewhere in the middle? (Hint: The truth almost always lies somewhere in the middle.)

Consider the following scenario. In January 1897, Boltzmann delivered a lecture to the Imperial Academy of Sciences in Vienna. Following the presentation, Mach stood up to address the group, not to ask a question about the content of Boltzmann's lecture (which was presumably very technical) but rather to declare: "I don't believe that atoms exist!"[9] End of discussion. No further comment. Mach might as well have said: *Sorry, Ludwig. I didn't listen to a thing you had to say. I don't even agree with your title slide.* He refused to debate Boltzmann on the merits of the science because he knew that it would be more effective to dismiss the work entirely. His message to Boltzmann was essentially: *Nothing you can say will ever convince me. I think that everything you've ever done in your life has been a waste of time.*

It's difficult to say (even now) how Boltzmann should have responded to Mach and his other critics. Hindsight is not always 20/20. (Aphorisms, be damned.) It's clear, however, that the prickly sense of exasperation that he sometimes conveyed in his later writings was almost certainly counterproductive. Consider this indignant gem: "Zermelo's paper shows that my writings have been misunderstood; nevertheless it pleases me for it seems to be the first indication that these writings have been paid any attention in Germany."[10] He had a knack for overlooking the obvious—that he might have made more allies (and fewer enemies) had he not allowed his frustrations to slip into his work unedited. He might, for example, have recognized that Ernst Zermelo, who held an influential position as Planck's assistant, was someone worth recruiting to (or at least not alienating from) his Team Boltzmann. But

Boltzmann was unaware of the impression that he sometimes made, and his response to criticism was not always commensurate with the opposition that he actually faced. At an 1895 debate in Lübeck, he upset his friend Ostwald by treating the event as if it were a blood sport. Boltzmann represented the atomic theory at the debate, while Ostwald stood in for Mach. Nearly everyone in attendance agreed that Boltzmann won the debate, but not without hurting Ostwald. One person described it as "the fight of the bull with the lithe swordsman."[11] Boltzmann, who was clearly the bull, made his arguments forcefully and with little subtlety, but he failed to realize that the debate was not a fight to the death (and that he, as the bull, did not need to gore the matador).

Boltzmann might have fared better personally had he not engaged his critics directly. He might have been better served had he simply ignored them and assumed a less public role. Granted, this strategy would not have worked in every circumstance. He needed to defend his work when challenged publicly, and he needed to respond when like-minded scientists corrected his actual physics. But he should not have felt compelled to submit his work for the approval of philosophers when there was simply too little common ground on which to hold a worthwhile debate. He never could have satisfied their burden of proof. Boltzmann was a scientist, and a scientist's goals are (seemingly) more modest than those of a philosopher. Scientists recognize patterns and make predictions. They ask questions and offer explanations. They never really prove anything, at least not by any philosophical standard. Proof requires a level of certainty and control that Mother Nature typically withholds from even her most ardent admirers. But few, if any, scientists seem to mind the terms of this agreement. Uncertainty bestows on their method a sense of skepticism, which is one of its most productive features. And the scientific method clearly produces results (which is a near-laughable understatement). Meanwhile, too much of philosophy sets for itself goals that seem almost deliberately unattainable or, in Mach's case, the exercise seems to produce nothing more than a list of thou shalt nots.

The irony (or is it tragedy?) of this debate over the early atomic theory is that the loser of the debate clearly benefited from it personally, while the winner plainly suffered. Mach was born at the right time and could have played no other role any better. He was a respectable scientist, but not an indispensable one. His shift toward philosophy seems to have been a wise career move. The same cannot be said of Boltzmann. Boltzmann was not just a respectable scientist; he was a rare event. As a philosopher, he was no Socrates (although he probably would have enjoyed the relaxed and roomy fit of a toga). Philosophical forays were a distraction for him, not a second calling. He should have remained focused on his physics. Had he not allowed himself to be drawn so directly into the fray, he would have seen Mach's program succumb to an inevitable lack of progress, and he would have seen the atomic theory supported by an overwhelming number of findings.

Einstein's theory of Brownian motion was in many ways the definitive evidence. It would convince all but the most recalcitrant of physicists that atoms really do exist. The theory originated in part in the work of Robert Brown, a Scottish botanist and microscopist who in 1827 observed that pollen grains move in a jittery sort of way when suspended in water. Other investigations found that these motions were ubiquitous, occurring not only with biological particles such as pollen grains but also in suspensions of inorganic particles such as coal. And, they found, the motions were persistent.

One could stabilize the platform on which the observations were made (the thought being that some perturbation might need to be suppressed), wait for long periods of time (to give those perturbations enough time to settle out), and still the motions would continue. The underlying mechanism, however, remained unclear until a twenty-something Einstein published a solution to the problem in 1905.

Einstein was not yet a science icon in 1905, although he was ready for launch that year. It was his annus mirabilis. In addition to the paper on Brownian motion, Einstein published three other papers in 1905 that were all history-of-the-world defining moments. One of those papers is considered to be a founding document of quantum mechanics, a second established the special theory of relativity, and a third gave us $E = mc^2$, the world's most famous equation. (Sorry, $S = k \log W$.) What's so remarkable about the annus mirabilis (from my limited perspective) is that the paper on Brownian motion is often overlooked, the motion of pollen grains simply failing to elicit the same level of respect as the other three topics. Yet it was the motion of pollen grains that provided what many physicists at the time considered to be the most convincing evidence for the existence of atoms.

The basic idea, which was not original to Einstein, was that the motion of pollen grains, which we can see under a microscope, is due to a constant bombardment by the surrounding water molecules, which we cannot see. The thinking was that although the water molecules are much smaller than the pollen grains, they are many times more abundant, which allows them to exert a collective effect on the larger particles. Einstein calculated the magnitude of this effect by using Maxwell-Boltzmann-like statistics. He knew that just as Boltzmann could not account for every atom in a gas, he too was unable to account for every water molecule in a suspension of pollen grains. But Einstein realized, as had his predecessors, that many useful properties could still be predicted without any detailed knowledge of the individual particles; a few general statistics were sufficient. One could, for example, calculate the likelihood that a slight excess of water molecules will collide with a pollen grain from its left-hand side. Imbalances of this sort are expected in a fluctuating bath of water molecules. When the pollen grain is assaulted from the left, it will be pushed to the right, but only for a brief moment because shortly after, the forces will inevitably turn in the opposite direction, with a slight excess of water molecules colliding with the right-hand side, causing the pollen grain to then bounce to the left. And this wayward back-and-forth will continue indefinitely with the pollen grain subjected to randomly changing forces that push it in a constantly changing direction.

The fact that these motions are random, however, does not imply that they are indecipherable. Hidden among the randomness are measurable quantities that the statistically minded Einstein was able to identify. He had studied Maxwell and Boltzmann, and he knew how positions and velocities are distributed across a large number of water molecules. He was also clever enough to see how that information, when applied to the motion of pollen grains, revealed something as elusive as the mass of an individual water molecule, which was a breakthrough. When the mass of a molecule became a known quantity, the atomic theory became an undeniable reality even for those segments of the physics community that had previously resisted it. (The chemists, meanwhile, probably just shrugged their shoulders and kept on working.)

All of this prompts us to ask why Boltzmann did not solve the problem of Brownian motion himself. Even Einstein admitted that he was fortunate to have inherited this problem. "It is puzzling," he noted, "that Boltzmann did not himself draw this most perspicuous consequence, since Boltzmann had laid the foundations for the whole subject."[12] It was a well-known problem, whose solution Boltzmann was better prepared than anyone else to find. He could have robbed Einstein of a key discovery, rendering the annus mirabilis perhaps slightly less mirabilis. But Boltzmann decided to engage Mach instead, on a subject of Mach's choosing, which distracted him from physics and distanced him from the discoveries of his day. It's possible that Boltzmann was not even aware of Einstein's discovery when he finally put his head in a noose. Ehrenfest's diary suggests that his mentor knew nothing about radioactivity,[13] which was another topic whose significance would have been difficult to overlook. Was Boltzmann too busy maladjusting to his new career as a philosopher? Or did he simply fall victim to one of the many reasons that older scientists sometimes fade from the forefront of discovery? Curiosity is for some a finite resource, and previous success can also act as an intellectual satiety factor. But productivity can also suffer when curiosity and drive remain intact as eyes and other organs on which the brain depends begin to fail. It sucks to get old, and Boltzmann did not age gracefully. Toward the end of his life, he was unable to manage his daily responsibilities, much less remain up-to-date on the emerging subjects of radioactivity and Brownian motion.

Still, the theory of Brownian motion feels like a lost opportunity for Boltzmann. He might not have been able (for any number of reasons) to solve the problem in 1905, but he could have solved the problem long before he fell into total disrepair. His own writings indicate that he understood the physical basis for the problem at least nine years before Einstein was credited with its solution. In a paper from 1896, Boltzmann wrote: "The observed motions of very small particles in a gas may be due to the circumstance that the pressure exerted on their surfaces by the gas is sometimes a little greater, sometimes a little smaller."[14] He was referring here to particles that are suspended in a gas, not a liquid, but the physics is very similar to that of the pollen grains in water. In both cases, the motion of something observable is due to collisions with something unobservable, and those unobservable somethings are readily described by the statistics of random fluctuations, a subject that Boltzmann helped to establish.

If only he had followed his remark with a calculation. He could have revealed the magnitude of the effect and the mass of those unobservable somethings. But instead he made the comment casually, as if it were obvious, and then proceeded in the paper to lecture Zermelo more generally on the role of fluctuations in physics. He failed to recognize how convincing an argument he could have made and left the topic open for an enterprising young Einstein to later claim. It was a missed opportunity for both physics and Boltzmann personally. Had he provided the demonstration himself (had he realized that his debate with Mach would be won by physics, not philosophy), he might have felt less threatened, less an outsider, and more secure in his many accomplishments. And he might have lived with himself long enough to eventually die of natural causes.

21 In Search of a Better Analogy

Imagine that you walk into an auditorium and find that everyone... Wait. We already covered that example and concluded that roughly equal numbers of atoms are found on each side of the room. But what if the room is the site of more dramatic goings-on? What if everyone is sitting on the left-hand side, but not because they're concerned in any way about the air? What if everyone has flocked to the left-hand side because bedbugs were found on the right-hand side of the room the day before? You start to itch just thinking about those little critters and wonder where they might have wandered since their original siting. You're no entomologist, but you know that bedbugs won't restrict themselves to the right-hand side of the room forever. You need a way to analyze the situation and the threat that it poses. So you decide to calculate the entropy.

The topic of bedbugs is obviously far removed from the study of gases and thermodynamics, but the concept of entropy is general enough to still prove useful. And almost every attempt to say something meaningful about entropy hinges at some point on whatever insights can be tied, often tenuously, to an overburdened analogy. The conventional script typically relies on references to either cracked eggs or perpetually

messy bedrooms. Our example here is motivated in form, if
not in detail, by a thought experiment that Boltzmann initiated and his student Ehrenfest later developed. To begin, we
need to specify the number of bugs, and four seems a modest
enough start. (We're nothing if not methodical in this book.)
We'll assume that all of the bedbugs are initially found on
the right-hand side of the room, but we'll set the bugs in
motion by randomly picking a number between 1 and 4 and
then moving the bug that was assigned to that number to the
opposite side of the room. We'll repeat this exercise several
times, each time picking a random number (1, 2, 3, or 4) and
moving the corresponding bug from wherever it currently
resides to the other side. Each time that we pick a number and move a bug, we'll add a new time point. And as the
time points accumulate, we'll generate a simulation of the
bedbugs' movements. You can, if you like, turn this thought
experiment into an active learning exercise by writing each
number on a piece of paper and then pulling a number out
of the hat of your choice. Just make sure that you return the
number to the hat before drawing the next time point and
keep track of the bedbugs as you go along. Table 21.1 is an
example of the sort of information that you might collect.
This particular simulation was stopped after ten numbers had
been drawn.

The table gives the positions of the individual bugs and
thereby represents a microstate-level description of the
thought experiment. The next step is to move toward a macrostate-level description. Imagine, therefore, what you would
see if you were not the director of the bedbug circus but simply a spectator, with no way of knowing how the bugs were
numbered or how the numbers were drawn. In that case, this
simulation would yield the observations in table 21.2.

Table 21.1
Simulation of ten bedbug jumps

Time point	Number drawn	Position of bedbug 1	Position of bedbug 2	Position of bedbug 3	Position of bedbug 4
0 (At the start, all four bedbugs are on the right.)		Right	Right	Right	Right
1	4 (note that bedbug 4 has moved from its initial position)	Right	Right	Right	Left
2	2	Right	Left	Right	Left
3	3	Right	Left	Left	Left
4	4 (note that a bedbug can make a round trip: moving from right to left, and then back to right)	Right	Left	Left	Right
5	1	Left	Left	Left	Right
6	3	Left	Left	Right	Right
7	4	Left	Left	Right	Left
8	1	Right	Left	Right	Left
9	2	Right	Right	Right	Left
10	1	Left	Right	Right	Left

Table 21.2
Macrostate-level description of the bedbug simulation

Time point	Observation	W
0	Zero bedbugs on the left, four bedbugs on the right	1
1	One bedbug on the left, three bedbugs on the right	4
2	Two bedbugs on the left, two bedbugs on the right	6
3	Three bedbugs on the left, one bedbug on the right	4
4	Two bedbugs on the left, two bedbugs on the right	6
5	Three bedbugs on the left, one bedbug on the right	4
6	Two bedbugs on the left, two bedbugs on the right	6
7	Three bedbugs on the left, one bedbug on the right	4
8	Two bedbugs on the left, two bedbugs on the right	6
9	One bedbug on the left, three bedbugs on the right	4
10	Two bedbugs on the left, two bedbugs on the right	6

We now no longer know the positions of the individual bedbugs, only the number of bugs found on either side. This macrostate-level description can also be represented graphically (figure 21.1), which helps us to identify the overall trend.

We see that the number of bugs on the right-hand side decreases from its initial value of 4 before then fluctuating around a value of 2, which is the value that's most frequently observed over the course of the simulation. Note, however, that this particular simulation is only one of many possibilities. It's a product of the random numbers that were drawn. If you were to generate your own simulation, it's very likely that you would observe a different set of outcomes. The overall trend, however, is likely to be the same, in that the number of bugs on the right will, after an initial decrease, eventually

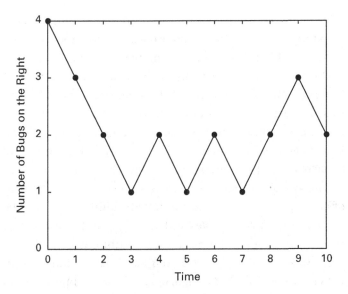

Figure 21.1
The number of bugs on the right over the course of the simulation

fluctuate around a value of 2. And if that trend is not immediately apparent, it can be found by simply adding new time
points to the simulation. (The example here is relatively
short in stopping after only ten random numbers.)

In order to see how this trend relates to the entropy, we
need to revisit table 21.2. There we find a column marked
with a W and learn (as we had suspected all along) that it was
never really about bedbugs. The bedbugs were just a stand-in
for atoms or other particles that occupy positions in space.
The bedbugs get us to macrostates with known multiplicities. The initial macrostate has a multiplicity of 1 because the
experiment is free of any uncertainty at the start. For a system of four bugs that are distributed across two sides of a
room, we know that there are 16 possible configurations or

microstates. (You can confirm this result by consulting figure 7.1.) But only one of those microstates places all four bedbugs on the right-hand side of the room, hence the multiplicity of 1. But when a number is drawn from the hat for the first time, the multiplicity increases to 4 because any one of the four numbers might be drawn, which would send any one of the four bedbugs to the left. Four random numbers lead to four possible microstates, which when lumped together give a macrostate whose multiplicity is 4 (table 21.3).

The key quantity to emerge from these numbers, be they of bedbugs or atoms, is the multiplicity. Multiplicities can be turned into entropies. We need only to take the logarithm. Here we've adopted the common logarithm, which refers to base 10. We have ten digits on our hands and ten digits with which we count (0, 1, 2, 3, 4, 5, 6, 7, 8, 9); our number system is a base 10 system. It seems reasonable, therefore, to adopt the common logarithm in a book (you're tired of hearing

Table 21.3
Possible outcomes for time point 1

Possible numbers drawn at time point 1	Position of bedbug 1	Position of bedbug 2	Position of bedbug 3	Position of bedbug 4	One macrostate
1	Left	Right	Right	Right	One bedbug on the left, three bedbugs on the right
2	Right	Left	Right	Right	
3	Right	Right	Left	Right	
4	Right	Right	Right	Left	

this by now) that aims to be simple. Please note, however, that should you be fortunate enough to encounter the subject of entropy elsewhere, you'll likely work with a different base. Physicists and chemists typically use the natural logarithm (to the base e) because it provides relatively simple results when the tools of calculus are required, while mathematicians and computer scientists often prefer the binary logarithm (to the base 2) because their basic unit of information, the bit, has two possible values. The choice of base therefore is largely a matter of convenience. Its only real consequence is the value that's assigned to the constant k. In order to ensure that the temperature scale is accurately aligned with the energy scale, the natural logarithm requires that $k = 1.381 \times 10^{-23}$ J/K. But for the purposes of this thought experiment, there really is no meaningful temperature. It would be possible, I suppose, to assign velocities to the bedbugs, along with the corresponding kinetic energies, which would introduce some sort of bedbug temperature for our consideration. But such a move would likely impose too severe a strain on the analogy. It seems reasonable, therefore, to limit our investigation to the positions of the bedbugs and to drop the constant k. The entropy is then simply $S = \log W$ (which is similar to the approach that's taken in many math and computer science applications, where again there's no need to consider a system temperature).

So let us calculate the entropy. Table 21.4 shows how the entropy changes over the course of the simulation.

When these data are plotted, the essential finding is revealed (figure 21.2).

At the start, the bedbugs are arranged in such a way that the entropy is especially low. In fact, the entropy starts at its lowest possible value of 0. This value is expected when the bedbugs are either all on the left or all on the right. Once the

Table 21.4
Entropy over the course of the simulation

Time point	W	$S = \log W$
0	1	0
1	4	0.60
2	6	0.78
3	4	0.60
4	6	0.78
5	4	0.60
6	6	0.78
7	4	0.60
8	6	0.78
9	4	0.60
10	6	0.78

bugs are liberated from this initial configuration, however, the entropy increases and then fluctuates near a maximum value, which for four bedbugs distributed across two sides of the room is 0.78.

What then are we to conclude? More specifically, what significance should we attach to the different entropy values? If you've not been paying attention, you might be forgiven, but you need to reengage now lest you miss the take-home message. Let's first consider the low-entropy initial conditions. The entropy is low when the multiplicity is low, and the multiplicity is low when the observed macrostate accounts for a small number of microstates. A macrostate with only one microstate is a limiting case, which gives the entropy its limiting value of 0. But the entropy does not remain low; it increases. And there we find our take-home message: as long as the bedbugs are provided with a mechanism by which they

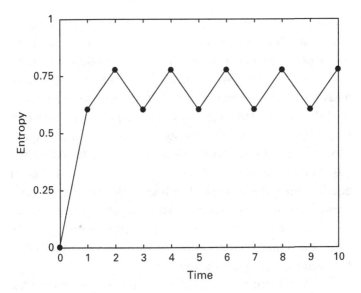

Figure 21.2
Entropy over the course of the simulation

can move freely and sample new configurations, the entropy increases, and the entropy continues to increase until the system arrives at its macrostate of highest multiplicity. At that point, the entropy reaches a maximum value, and the system is considered to be at equilibrium. The entropy therefore acts as a sort of progress variable whose rise traces a path toward equilibrium.

But what is equilibrium (aside from the state of maximum entropy)? The word seems to carry with it a positive connotation. It's something that we might achieve, something aspirational. It conveys a sense of balance. If we eat the right food, work hard at our jobs (but not too hard), cultivate fashionable hobbies and commendable habits, we might one day reach a state of equilibrium in our lives (which we can then proudly display on social media). But such a standard seems almost unattainable. (I speak, of course, for only my train wreck of a self.) Our personal lives are complicated in ways that these simpler systems are obviously not. Equilibrium in a gas is something easily achieved. It happens spontaneously. And therein lies the utility of the concept. If we know that something occurs spontaneously, we know what to expect. More specifically, we know what to expect *in the future*. If a gas, in its present state, is not at equilibrium, it will be at equilibrium in the future, and we'll know that the gas is at equilibrium because its entropy will have stopped increasing. The increase in entropy therefore serves a critical role in establishing an unambiguous arrow of time—a clear distinction between the past, present, and future.

This statement sits somewhere on the border between the trivial and profound, although its profundity may require some further justification. Let us ask ourselves, therefore, how it is that we make sense of the world. One could argue that the world makes little sense (especially in the realm of human affairs). But consider how much worse it would be if not for spontaneous processes. Gases diffuse (filling their containers) and ice cubes melt (making puddles of water), and our world is a somewhat comprehensible place because these systems spontaneously approach well-known equilibrium states. If that were not the case (if these outcomes were not so obvious and utterly predictable), we would struggle to make sense of anything. Our capacity to reason would serve no useful purpose because Mother Nature would no longer be a rational thing in which to place our trust. Reason is built on the assumption that spontaneous processes happen spontaneously and always meet our expectations. And now that we've tied our argument to something so grandiose as our capacity to reason and the arrow of time, we seem to have once again turned to philosophy, which may have been unavoidable (or may be an indication that I'm a hypocrite and favor the topic only when it suits my purposes).

In any case—if we're going to conclude that increasing entropy plays such a fundamental role—we should then test the idea more thoroughly. We should, at the very least, include a great many more time points in our simulation. Figure 21.3 shows the results that were obtained from a simulation that was extended to time point 100.

The most striking feature of the longer simulation is the relative instability of the equilibrium state. If we identify the equilibrium state as one of maximum entropy, we should then expect the maximum entropy value to persist once it

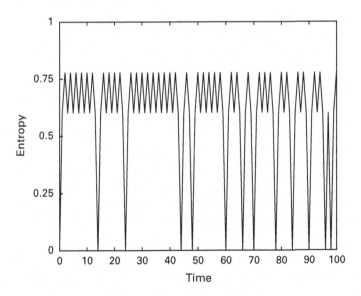

Figure 21.3
Entropy over the course of an extended simulation

is reached. But in this example, the entropy actually returns to its initial value multiple times, indicating that equilibrium maintains a rather weak hold on the system. The system tends toward equilibrium but makes frequent departures from it, and those departures are severe in the sense that the system actually revisits its lowest entropy initial state. The arrow of time appears to be wavering, which makes for flimsy predictions. If it is to be useful, the arrow needs to be steadfast, and equilibrium needs to exert more dictatorial control over the system. Fortunately, we know how to fix this problem. Experience has taught us that big numbers often suppress irregularities. Longer simulations therefore may simply require bigger numbers.

22 Equilibrium You Can Count On

A minor infestation is enough to reveal the essential trend. We can see the trend clearly with as few as 100 bugs, although a relatively large number of time points are then needed in order to reach equilibrium. So let's start with 100 bedbugs, all of which are initially on the right-hand side, but then allow the bugs to jump 1,000 times. Before starting the simulation, however, consider how tedious a task this would be if you were to attempt it by hand. You would need to write numbers on 100 pieces of paper, pick a number 1,000 times, and accurately record the results every step of the way. This simulation practically prohibits the hands-on approach that we previously endorsed. A computer is much better equipped for this task than you or me or anyone else whose time is worth anything at all. And a computer can perform multiple simulations with the investment of little additional time. Although the details are expected to vary among simulations, the overall trend, which is illustrated in the entropy curve in figure 22.1, is not.

The entropy starts at 0 and then quickly rises, reaching its maximum value within the first 150 time points. Following that early transition, a stable equilibrium is then observed. Time points continue to accumulate, and bedbugs continue

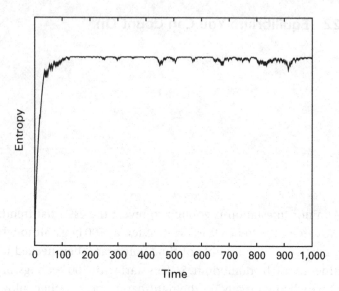

Figure 22.1
Entropy over the course of a longer simulation,
with a larger number of bugs

to jump back and forth. But the entropy deviates relatively little from its maximum value once equilibrium takes hold of the system. The entropy does continue to fluctuate, however, because the system is rarely observed with exactly 50 bugs on the left and 50 bugs on the right. Such a perfectly balanced scenario would demand too much of Mother Nature; we need to allow her a small excess of bugs on either side. The plot in figure 22.2 indicates what we should in fact expect. It shows how the number of bugs on the right changes over the course of the simulation.

We see that the number of bugs on the right is seldom a spot-on 50, but it fluctuates around 50, staying mostly between the values of 40 and 60. The entropy therefore is

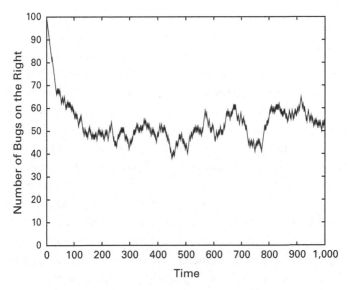

Figure 22.2
The number of bugs on the right over the course of the simulation

rarely its maximum value, but it never strays far from it. Temporary decreases are observed, but those decreases have relatively little effect on the overall shape of the entropy curve except to impose a tiny bit of noise on top of the plateau that is reached. That noise is small in comparison to the size of the entropy increase that occurs early in the simulation (i.e., as the system approaches equilibrium). These observations reveal that while the system is permitted brief excursions from its maximum entropy value, it's kept on a relatively short leash. Bigger numbers therefore do provide a more stable equilibrium (as I had predicted). But bigger numbers also suggest a need to expand our view of the equilibrium state—to include not only the maximum entropy value but also the values in its vicinity. Equilibrium therefore is not the serene destination that we might have once envisioned, but something a bit noisier instead.

If this more generous view of the equilibrium state fails to meet your expectations, you may then sympathize with the nineteenth-century response to Boltzmann's entropy. It is a function that clearly fluctuates at equilibrium, which was not considered a welcome development. Consider the circumstances. Previous descriptions of entropy had been based on the study of mechanical objects—things like steam engines, whose thermodynamic properties could be understood without recourse to statistics. The goal of those early studies was to improve the efficiency of engines—to maximize the amount of work that could be performed for a given amount of heat input. The entropy that emerged from those engineering applications was considered an inviolable concept. It increased as a system approached equilibrium and then maintained a constant value once equilibrium had been reached, which was exactly what was considered necessary in order to establish an unambiguous arrow of time. This early interpretation also exemplified the sort of unfailing reliability that one might expect from a physical law—in this case, the second law of thermodynamics. Although the second law was revised and reinterpreted many times throughout the nineteenth century, a common theme that persisted across its many reincarnations was the notion that entropy did not decrease. It would increase if the system was not yet at equilibrium or remain constant if the system was already at equilibrium, but it did not decrease. So said the second law.

Boltzmann's entropy, however, seemed to contradict that notion, which led many of his fellow scientists to suspect that he had ruined the whole thing. His entropy probed much deeper than had previous accounts, but there was an unwelcome price to be paid for the progress that he had made. His use of statistics had introduced fluctuations, and a quantity that fluctuates sometimes goes up and other times goes down (by definition). Entropy had become a shaky quantity in Boltzmann's hands. It was no longer something that could be trusted, and the second law was suddenly suspect. What good is a law that holds only some of the time? And why would anyone agree to compromise something so fundamental? Would it not make more sense to dismiss Boltzmann's findings than to amend the second law so drastically?

The physics community faced a moment of disillusionment in confronting these questions. And though some of its more well-informed members recognized that the problem readily resolved itself when bigger numbers were considered, many others argued that the arrow of time should be free from all threats. If presented with the results of our simulation (figure 22.3), they would have insisted that the entropy curve be airbrushed of its pesky noise.

If entropy can fluctuate, they asked, what would prevent one of those fluctuations from turning into something problematic? Perhaps it would not forever fluctuate in staccato-like fashion near its plateau. Maybe it would one day dive or simply drift downward over time, returning to its zero value at some not-yet-sampled time point.

These concerns crystallized in the work of Henri Poincaré whose recurrence theorem proved that the system will in fact return to its initial state. You simply need to wait long enough. At some point in the future, the entropy will decrease (and significantly) as all 100 bedbugs return to the right-hand side of the room. And if you're especially patient, the recurrence theorem also predicts that you'll see an ice cube melt and later refreeze, even though the thermostat maintains a cozy

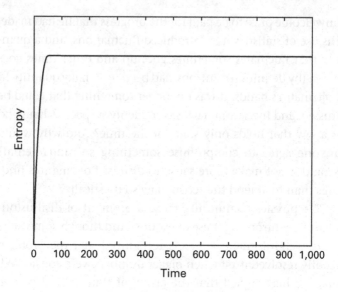

Figure 22.3
A simulation without noise

72 °F (= 22 °C = 295 K). The recurrence theorem, however, does not tell us how long you might have to wait in order to observe these events. It is a mathematician's construct. It tells us what's possible (even necessary) in theory, but makes no attempt to identify what's reasonable to expect in practice. It is, for us, a cautionary tale—that mathematics, while an obvious source of inspiration in physics, is not an obvious indicator of what to expect in the natural world. Why? Because a mathematician's theorems are unencumbered by external reality in a way that a physicist's theories are not. A mathematician can prove that an event will occur in a finite amount of time and feel that he or she then adequately grasps the problem. But a physicist needs to contend with additional constraints. A physicist needs to discriminate between events that have

merely a finite chance of occurrence and those that are actually foreseeable in the lifetime of the universe. The physicist's objective therefore is fundamentally different from that of the mathematician, and the physics community has good reason not to yield too much of its authority to its mathematical brethren.

Many of Boltzmann's colleagues, however, were quick to defer to Poincaré, who was widely regarded as an exceptionally smart man (with a cool French name). The fact that Poincaré had found apparent fault in Boltzmann's entropy led many to assume that it was suspect, and for those who wanted to challenge Boltzmann, the recurrence theorem was a source of new ammunition. It conferred a sense of mathematical respectability on their suspicions and gave them talking points that were more likely to prompt reflexive nods of agreement rather than scrutiny. Planck's assistant Ernst Zermelo was especially quick to capitalize on the Frenchman's findings. He initiated an exchange of papers with Boltzmann in which the recurrence theorem was volleyed back and forth between the two scientists. In those papers, Zermelo insisted that the recurrence theorem precluded a mechanical view of the second law. He contended that any reference to atoms or other particles bouncing around, their positions and velocities ever changing, be discarded for the sake of the second law. One could have one or the other but not both, thought Zermelo, and he much preferred to preserve the sanctity of the second law. If the microscopic constituents of matter behaved mechanically, he argued, then a system would not approach equilibrium irreversibly. The room temperature puddle of water would eventually reconfigure itself into an ice cube, leaving the observer with no apparent way to discriminate between past and future.

Boltzmann of course promoted a different view. He respected the recurrence theorem as a work of mathematics, but maintained that it had few physical implications due to the inconceivable timescales that it implied. No one, he explained, has ever observed an ice cube melt and later refreeze at room temperature because a round trip of that sort (from a low-entropy state to a high-entropy state and then back to the low-entropy state) would require a period of time that renders the age of the universe insignificant. Nor has anyone ever witnessed a gas expand to fill a room and later reconsolidate itself in an isolated corner of that room, again because the time needed for a system of that size to return to its initial low-entropy state is ludicrous, which Boltzmann clearly demonstrated to Zermelo. He described the matter to Zermelo this way: "Thus when Zermelo concludes, from the theoretical fact that the initial states in a gas must recur—without having calculated how long a time this will take … he is just like a dice player who has calculated that the probability of a sequence of 1000 one's is not zero, and then concludes that his dice must be loaded since he has not yet observed such a sequence!"[1]

For Boltzmann, it was a matter of simple common sense that would apply equally well to our system of bedbugs. Imagine, for example, that the system has already reached equilibrium, with 50 bugs on either side. Now it's certainly possible that over the course of an additional 50 jumps, all 50 of the bugs on the left will return to the right-hand side of the room. But it would require a significant decrease in entropy and is therefore highly unlikely. How unlikely? We can calculate the odds and find out. At the start of these 50 additional jumps, 50 of the 100 bugs are on the left. We need to pick one of them (in order to begin the process of moving the bugs back to the right). Since there are 100 possibilities, 50 of which would give the desired result, the probability is 50/100 (50 out of 100). Not such bad odds, but this jump is only the first of 50 that have to go a particular way. In the next jump, we have 49 bugs on the left and again need to pick one of them. The total number of bugs is the same as before, so the probability of this second event is 49/100. And with each subsequent jump, we have fewer and fewer bugs on the left, which means that it's less and less likely that we'll pick one of them. The probabilities therefore drop from 50/100 to 49/100 to 48/100 on down to 1/100. And when we string all 50 of these events together, we find that the probability of everything going as planned (of sending all of the bugs back to the right in 50 jumps) is

$$\frac{50}{100} \times \frac{49}{100} \times \frac{48}{100} \times ... \times \frac{1}{100}$$

If you're familiar with the factorial function, you'll recognize that this expression can be written more compactly as $50!/100^{50}$, which may or may not make clear the fact that the numerator, while itself a big number, is many times smaller than the denominator. And when a smaller number is divided by a bigger number, we get a small number—in this case, a very small number, one that begins with 0.000000... and continues with many more 0s before arriving at something other than a 0. So, in effect, the number is something very close to 0, which allows us to safely conclude that the system has very little chance of realizing this low-probability sequence of events. Alternatively, the time that's needed for the system to return to its initial low-entropy state is a very long time indeed. And here we've considered a system of only 100 bugs. The probability would be even smaller if we were to consider a larger system.

But Zermelo was unswayed by these arguments and claimed that Boltzmann's reply was an admission of defeat. For Zermelo, an entropy decrease of any magnitude—equilibrium fluctuation or not—was a big deal, and any recognition of the fact that entropy can decrease discredited the way in which the entropy had been formulated. He simply expected more from an entropy function than what $S = k \log W$ could provide. Or, one might argue, he expected less—an entropy function that was insensitive to fluctuations among the microscopic constituents of matter. He insisted that entropy be still, not just stable, at equilibrium, and he knew that Boltzmann was somewhat vulnerable on this matter.

Although Boltzmann had been remarkably consistent in his exchange of papers with Zermelo, his overall record on the topic was actually riddled with contradictions. And in some cases, he seemed almost unaware of his own work. He displayed particular ambivalence about an earlier achievement called his H-theorem. The H-theorem was Boltzmann's first major success. It established his reputation as a world-class scientist and is admired still today as a monumental piece of mathematical physics. Yet the H-theorem contained a critical assumption (called the *Stosszahlansatz*) that undermined its main conclusion, which (prepare to be disappointed) was that the path from a low- to a high-entropy state was a one-way street. Entropy did not fluctuate under the H-theorem; it proceeded in a single direction. And so Boltzmann was on the record (quite unambiguously) in support of a more traditional view of the second law (the one that he eventually helped to replace).

You might at this point wish that the story would simply cooperate and adhere to our oft-stated goal that we keep things simple. Or if the historical record really is that complicated, you might prefer to be spared from all but the most essential of the remaining details. But if we're going to maintain a modicum of objectivity here, we need to acknowledge the fact that Boltzmann was not always a good advocate for $S = k \log W$, due largely to the fact that it came at a personal cost for him. He was not able to have $S = k \log W$ and still hold on to his H-theorem. The differences between the two theories were irrefutable. Boltzmann therefore needed to renounce his earlier achievement or else drastically reimagine it. But he was justifiably proud of the earlier work and sentimentally attached to it. If he had been somewhat less proud or less sentimental (an admittedly unfair expectation), he might have been able to assess his own research trajectory and recognize that he had explored a more traditional view of the second law before arriving at its more nuanced version—that he had needed to work within the confines of a tradition before knowing where that tradition might fail him. But Boltzmann never fully appreciated his own intellectual development on this subject, and he never produced a consistent summary of his own work. Even his *Lectures on Gas Theory*, which should have played that role, needs to be read selectively, which is a problem, because an uninformed reader is not likely to know which passages to select and a well-informed reader may not find it any easier. There was perhaps no better-informed reader of Boltzmann than Einstein, but even Einstein found Boltzmann's writings to be a source of frustration as well as inspiration, telling one of his students, "Boltzmann's work is not easy to read. There are great physicists who have not understood it."[2]

One of those great physicists was likely Niels Bohr, another founding father of quantum mechanics, who mentored many of the most influential scientists of the twentieth century.

Boltzmann's standing among a whole generation of physicists was likely diminished by the fact that Bohr did not consider Boltzmann's papers to be essential reading.[3] But Bohr may have spared his students a considerable number of headaches by diverting them from those papers. Boltzmann's attempts to own both sides of the entropy issue created a historical thicket that has never been easy to traverse. Does the second law prohibit a decrease in entropy or not? Boltzmann knew the answer. He knew that the second law offered likelihoods, not absolutes, but he never delivered a consistent view on the matter, and so relatively few scientists have ever bothered to read the original Boltzmann. Most scientists have relied on secondary sources instead.

Secondary sources have proven to be not only more reliable than Boltzmann's writings, but also free of many of the quirks that are particular to his writing style. His skill in the lecture hall (as a deft expositor) did not transfer to the written page. He rarely used consistent nomenclature, his assumptions were not always clearly stated, and he sometimes changed his assumptions without first notifying his readers. In fact, his papers often read as if they were sampled from his own personal lab notebooks, which is a style not entirely without merit. His use of specific model systems, for example, often helped to make abstract concepts more concrete. His 1877 paper benefited, in particular, from the case study of the seven atoms with a total energy of 7. The significance of the Boltzmann distribution may have eluded some of his readers were it not for that example. In that same paper, Boltzmann was even kind enough to include (or too lazy to exclude) a calculation that demonstrated how to avoid what he anticipated would be a common mistake (perhaps because he had already made that mistake himself). But his rambling style did not compare favorably overall with the rest of the scientific literature, for the simple reason that he cared not enough for appearances. He freely admitted that he did not strive for elegance in his work. Elegance, he joked, was "for the tailor and the shoemaker."[4] And photographs suggest that he rarely consulted his own tailor or shoemaker (or barber)—which might not have been much of a problem for him personally. But his writing was a different matter, and it suffered real consequences from an inattention to style.

The problem does not seem to have originated in his methods. His thinking was not in any way disorganized, nor was his path to discovery any messier than that of other scientists. His approach, like that of many other scientists, was a typical shotgun approach—with unsuccessful attempts and theoretical dead-ends, broken assumptions, and the occasional inconsistency. But Boltzmann was unusually willing to put that process on display, and he somehow resisted the nearly universal urge among scientists to manufacture a polished story. Or perhaps he simply never felt that particular urge. In any case, he failed to follow the standard protocol, which all but demands that a scientist tell his or her story with the benefit of hindsight. According to the standard protocol, a scientist reveals his or her findings not in the order that they were found, but in an order that is thought to be most coherent and convincing. Unsuccessful attempts and theoretical dead-ends are omitted (for the sake of clarity, of course), and scientific history is often rewritten before it is even recorded, which is often a useful deceit. But it was not Boltzmann's style. David Lindley, a popular science writer who has written extensively about nineteenth-century physics, has suggested that Boltzmann wrote with all the elegance of a bulldozer, which seems the perfect analogy.[5] A bulldozer is obviously a blunt instrument, but no one would deny the bulldozer its usefulness (or its possible charms). If entropy was an object in need of finding, then that object was deeply buried at an ill-defined location. Lots of earth needed to be moved, and a more delicate instrument might not have been able to accomplish the task. A bulldozer was needed, and a bulldozer was ultimately successful. But the bulldozer drastically changed the landscape in the course of its discovery, leaving future generations the task of reconstructing a path that seemingly no longer exists.

23 A Rigorous Examination of the Obvious

Imagine that you walk into an auditorium (yes, I apparently have a limited imagination) and note nothing odd in the way that everyone is distributed about the room. Roughly equal numbers of people are sitting on the left and on the right. The people on the right, however, are all shivering, while the people on the left are sweating and beginning to exhibit various stages of undress. You wander around the room a bit in order to investigate and confirm that the right-hand side of the room is in fact much colder than the left, but you see no obvious reason for the discrepancy. There's no window open on either side of the room, nor does there appear to be any problem with the vents to the room's heating and air-conditioning system. It seems that the temperatures of the two sides have simply diverged from one another, and you suspect that you know why: you suspect that the atoms on the left have accumulated more kinetic energy than the atoms on the right, which the people in the room now experience as a temperature difference.

You quickly realize, however, that you've never observed an outcome of this sort in the past, and such a result would seem to constitute an actual weather pattern inside the room, which is ridiculous. But how ridiculous? We now have the tools to answer that question. We can determine the likelihood that the atoms on one side of a room would spontaneously acquire more kinetic energy than the atoms on the other side. We can start simple and keep it simple. And in this case, we don't even need to calculate the entropy of the auditorium. We can calculate the multiplicity of a much smaller room instead.

So let's say that there are four atoms on the left and four atoms on the right. And let's allow each atom to have a kinetic energy of either 0 or 1. Such a system is called a two-state system because there are only two allowed energy values for each atom. A two-state system provides a convenient starting point from which to investigate the direction of heat flow, our goal here. In order to achieve that goal, let's assign a temperature to each side of the room by allotting each side some amount of energy. Given the fact that the left-hand side is considered to be hotter than the left, we'll assign an energy of 3 to the left and 1 to the right. We can then count the number of ways that this energy can be distributed across the atoms on each side. In the case of the left, there are four atoms that share an energy of 3. If each atom can have an energy of either 0 or 1, then one of the four atoms must have 0 energy and each of the other three atoms must have an energy of 1. Since any one of the four atoms could be assigned the 0 value, there are then four microstates associated with the left (table 23.1).

A similar reasoning applies to the right, although there's less energy to be shared among its four atoms. The four atoms on the right need to share an energy of 1. This constraint dictates that one of the atoms must have an energy of 1, which leaves 0 energy available for the other three. In that case, there are again four microstates because any one of the four atoms could be assigned the value of 1 (table 23.2).

Table 23.1
Microstates for the left-hand side of the room

Microstate	Energy of atom 1	Energy of atom 2	Energy of atom 3	Energy of atom 4	Allotted energy (just to check)
1	0	1	1	1	$0+1+1+1=3$
2	1	0	1	1	$1+0+1+1=3$
3	1	1	0	1	$1+1+0+1=3$
4	1	1	1	0	$1+1+1+0=3$

Table 23.2
Microstates for the right-hand side of the room

Microstate	Energy of atom 1	Energy of atom 2	Energy of atom 3	Energy of atom 4	Allotted energy
1	1	0	0	0	$1+0+0+0=1$
2	0	1	0	0	$0+1+0+0=1$
3	0	0	1	0	$0+0+1+0=1$
4	0	0	0	1	$0+0+0+1=1$

What then is the total number of microstates across the entire room? In order to answer that question, we need to recognize the fact that each of the microstates on the left could be paired with any one of the four microstates that are possible on the right (figure 23.1).

The total number of microstates across the entire room therefore is $4 \times 4 = 16$. But a larger number of microstates is possible still if we allow the second law to now exert its influence.

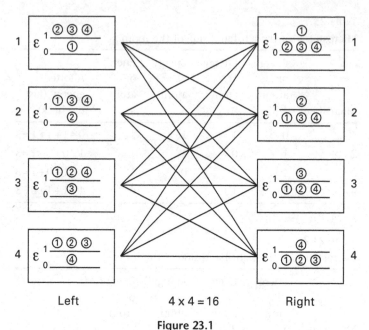

Figure 23.1
The total number of microstates across the entire room,
obtained by pairing each microstate on the left with each
microstate on the right

Imagine then what would happen if the excess energy on
the left were somehow transferred to the energy-deficient
right. The two sides of the room could then be allotted the
same amount of energy, which would be consistent with
our everyday experience in which the two sides of an otherwise ordinary room have the same temperature. In this more
equitable scenario, each side of the room would be allotted
an energy of 2. Does this more equitable scenario generate
a larger number of microstates? We can answer that question by first listing all of the ways that an energy of 2 can be

Table 23.3

Microstates for the left-hand side after the redistribution of energy

Microstate	Energy of atom 1	Energy of atom 2	Energy of atom 3	Energy of atom 4	Allotted energy
1	0	0	1	1	$0+0+1+1=2$
2	0	1	0	1	$0+1+0+1=2$
3	1	0	0	1	$1+0+0+1=2$
4	0	1	1	0	$0+1+1+0=2$
5	1	0	1	0	$1+0+1+0=2$
6	1	1	0	0	$1+1+0+0=2$

distributed across the four atoms on the left-hand side of the room (table 23.3).

We do not need a separate table for the right-hand side because it would be identical to the table generated here for the left. The two sides of the room now have both the same number of atoms and the same energy allotted to them. We simply need to recognize that each of the six microstates listed in the table for the left could be paired with any one of six microstates also available on the right. The total number of microstates for this equal-energy scenario therefore is $6 \times 6 = 36$ (figure 23.2).

Let us review. We started with a scenario in which one side of the room (the left) had more energy available to it than the other (the right). In that sense, the left was regarded as being hotter than the right. But we allowed the excess energy on the left to be transferred to the right, which then led to equal energies. Given the fact that the two sides of the room had the same number of atoms, equal energies in this

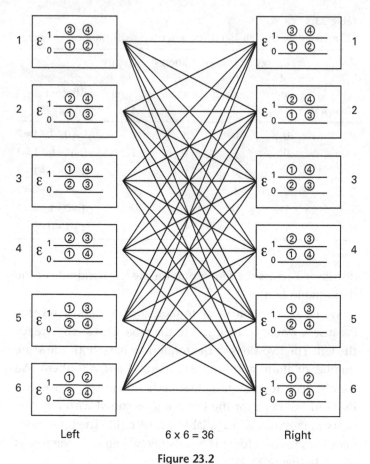

Figure 23.2
The total number of microstates across the entire room,
after the redistribution of energy

case would also imply equal temperatures. How then did the total number of microstates change in the process? The room shifted from a macrostate whose multiplicity was 16 to one whose multiplicity was 36. The total number of microstates therefore increased from 16 to 36. And more microstates means more likely (because Mother Nature freely samples the microstates that are available to her). The equal-energy state at which the system arrived therefore would be more than twice as likely as the unequal energy state where it began (because $36/16 = 2.25$).

Although the magnitude of this difference is not overwhelming, the number of atoms in this particular case is quite small. And we've noted several times before that unlikely scenarios become even less likely as the number of atoms increases. If, for example, we increased all of the quantities in our previous example by a factor of 10, the number of atoms on each side would increase from 4 to 40; the energy allotted to the left would increase from an initial value of 3 to 30; and the energy allotted to the right would increase from an initial value of 1 to 10. If this system were then allowed to redistribute its energy so that each side of the room had an allotted energy of 20, the total number of microstates across the entire room would increase by much more than a factor of 2.25 (as it had before). The total number of microstates would increase instead by a factor of over 26,445. By increasing the size of the system (even by a relatively small amount), an energy difference across the room becomes significantly less likely. Why? Because the macrostate in which the two sides have equal energy is represented by a far greater number of microstates. And if Mother Nature freely samples the microstates that are available to her, she'll arrive at that equal-energy macrostate many more times than any other.

So what should we expect in our auditorium with its two distinct climates? Can we provide any relief to the victims of this thought experiment? The people on the left would prefer not to sweat, and the people on the right would like to feel their toes again. What good news can we provide them? We can assure the people in the auditorium that a temperature difference of any noticeable effect is highly unlikely to have arisen spontaneously. In fact, the scenario, as we described it, is another one of those cases where the probability is not exactly 0 but is so close to 0 that for all practical purposes, it is inconceivable. The temperature difference that we observed on entering the room was not a spontaneous fluctuation. It would have required some sort of intervention. The room did not arrive at such an unlikely state on its own. Something must have happened before we entered the room, and that something was most certainly not a matter of statistical physics. So let us shift our attention from the question of how the room was initially prepared to its subsequent fate.

The transfer of energy in this thought experiment occurred in a particular direction. Energy was transferred from the hot side of the room to the cold side. The justification for this outcome was that it led to an increase in the number of microstates. There were simply more ways to distribute the total energy across the entire room when that energy was shared equally by the two sides. This finding suggests that if Mother Nature finds herself in an unlikely macrostate (for reasons that are perhaps outside her control), she will find her way to a more likely macrostate through the transfer of energy. But that transfer occurs in a specific direction—from hot to cold. So says the second law, and so said an important figure in the history of thermodynamics whom we have not yet recognized, Rudolf Clausius.

Clausius actually coined the term *entropy* and was one of the giants on whose shoulders Boltzmann later stood. He was responsible for several of the earliest formulations of the second law, but for our purposes here, we'll focus on his statement that "heat can never pass from a colder to a warmer body without some other change, connected therewith, occurring at the same time."[1] Clausius's view of the second law therefore focused on the direction of heat flow. It was a practical view motivated by engineering applications. And from that perspective, one can readily see why the second law would have been considered an inviolable concept. Heat flows through a steam engine in a predictable direction—from a heat source to a cold sink. And other macroscopic objects behave similarly. I previously mentioned that an ice cube inevitably melts when it sits at room temperature, but we're now much better prepared to recognize the reason for the transition. The ice cube and the room in which it sits are initially at different temperatures, a driving force for the transfer of energy. The second law according to Clausius predicts that heat will flow from the higher-temperature object to the lower-temperature object. Energy will be transferred, therefore, from the room to the ice cube. But once the ice cube melts into a room-temperature puddle of water, it is a relatively stable thing. It will evaporate over longer timescales (a process that leads to even higher entropy), but the puddle will not reconfigure itself into an ice cube at room temperature. Why? Because there's no driving force for such an unlikely event. The initial melting process was a one-way transition.

If we were to stop and take questions at this point, one might ask why such an elaborate explanation is needed for such a trivial outcome. The fate of an ice cube is obvious, one could argue. And one might challenge the significance of

Boltzmann's contributions if some other guy named Clausius actually formalized the direction of heat flow and thereby introduced the concept of entropy. These questions are not unreasonable, so let us address them, focusing first on the historical significance of Boltzmann.

It is true that our understanding of thermodynamics developed gradually, in fits and starts and with all of the virtues and vices of a poorly organized team effort. Its development was not like that of classical mechanics, where Newton was clearly the team captain and its founding document, the *Principia*, was an almost singular landmark. Thermodynamics was a much messier enterprise. But Boltzmann's role in that enterprise was critical because he provided the overall framework with an underlying mechanism. His adherence to the atomic theory and his steadfast use of statistics were modern in a way that his predecessors' methods were not. He took what was obvious—that an ice cube melts—and provided a justification for that outcome that was rigorous. He probed more deeply and revealed the truer nature of things. Why does an ice cube melt at room temperature? Because that change leads to a larger number of microstates and a higher multiplicity and a greater entropy.

Yet why does this subject merit such intense study, one might ask, given the fact that several of its outcomes appear to be so obvious? And why do we celebrate someone who made this particular subject his life's work? Although questions of this sort can never be answered to the satisfaction of all, they can be adequately addressed for some by our oft-repeated admission—that the goal of this book has always been to keep things simple. Yes, the fate of an ice cube may appear to be obvious, but an ice cube is a relatively simple thing. Other objects to which the second law applies are not so simple.

Even the nineteenth-century steam engine is complicated in a multitude of ways that are not addressed by this book. But fret not, my friends; there is good news (hallelujah!). If you are tired of my mantra—if you no longer want to keep things simple—you can seek out all of the many complex things that now await your consideration. Because if you go looking for entropy, you will find it—in applications that range from biochemistry to cryptography to economics to consciousness to the large-scale structure of the universe. But before we adjourn, let us say goodbye to Boltzmann. And before we say goodbye, let us accompany him on a farewell tour.

24 The Farewell Tour

He was anxious but not fearful, and restless but not adrift. He lived his life to the proverbial fullest (until of course the moment that he ended it). He traveled extensively, making two transatlantic trips in the last two years of his life—at a time when and in a condition where a more timid soul would have clung to the comforts of home. The last trip he documented in a short travel memoir that he titled *A German Professor's Journey into Eldorado*, which is a surprisingly funny sketch (funny in a wry, grumpy-old-man sort of way). He described his trip's departure this way:

The first part of my journey took place in a hurry and will also be told in a hurry. On June 8th I attended the Thursday meeting of the Vienna Academy of Sciences as usual. As we were leaving, a colleague noticed that I was not turning towards the Bäckerstraße as expected, but towards the Stubenring, and he asked where I was going. To San Francisco, I answered laconically. In the restaurant of the Nordwestbahnhof I consumed a leisurely meal of tender roast pork, cabbage and potatoes and drank a few glasses of beer. My memory for figures, otherwise tolerably accurate, always lets me down when I am counting beer glasses.[1]

Before attempting to offer anything here that might resemble a serious comment, first note that all of the following words presumably refer to real places: Bäckerstraße, Stubenring, and Nordwestbahnhof. (Where else would you expect to find a guy named Boltzmann?) You should also know that Boltzmann's account of his departure represents the man more faithfully than any additional analysis that I'll be able to provide here. By his own account, he seems not to have taken himself too seriously. A trip from Vienna to San Francisco would not have been a trivial undertaking in 1905, especially for an unaccompanied old man in deteriorating health. Yet Boltzmann portrayed the trip as just another day in the life (along with an implied wink to his reader). Note that he was also quick to mention his fondness for food and drink in these first few paragraphs. And in fact much of the story that follows his oft-quoted introduction is an amusing combination of food diary and personal medical record, from which we quickly learn that: he didn't like the water in California; he didn't like the air; and he couldn't self-medicate (at least initially) because his destination was the dry town of Berkeley. He was able to solve the last of these three problems, but only after an uncomfortable conversation with an American colleague that proceeded as follows: "He [the American colleague] looked about anxiously in case someone was listening, sized me up to see if he could really trust me and eventually came out with the name of an excellent shop selling Californian wine in Oakland. I managed to smuggle in a whole battery of wine bottles and from then on the road to Oakland became very familiar."[2]

We might conclude from this story that had he not been such a successful scientist, Boltzmann might have had a successful career instead as a California bootlegger, although he probably lacked the necessary ruthlessness for the job. Ruthless he was not. But he was perturbed by the temperance laws in the United States, which he rightly regarded as hypocritical. He joked, in describing his exchange with the American colleague, that the man's response would have been more appropriate had the inquiry concerned the whereabouts of a San Francisco prostitute and not some Oakland wine shop. It is also somewhat amusing to note that it had not escaped Boltzmann's attention that one could find in San Francisco "girls there with the motto: 'Give me money, I give you honey.'"[3]

Boltzmann likely avoided the exchange of any of his money for some honey, but he did manage to stumble into a few less decadent misadventures. Early in his trip, for example, he was invited to a dinner party that was hosted by Mrs. Phoebe Hearst—principal benefactor of the University of California at Berkeley, financial sponsor of his trip, and mother of newspaper magnate William Randolph Hearst. Mrs. Hearst was fabulously wealthy. If she had fancied it, she could have served her dinner party salads made of freshly chopped $100 bills. Yet despite her enormous wealth, Mrs. Hearst likely experienced some philanthropic version of buyer's remorse after meeting Boltzmann because her guest did little if anything to express his gratitude, which seems to have been lacking. He was, by his own account, a less-than-gracious guest:

At table I sat on Mrs. Hearst's right, as I was the only European present. The first course was blackberries. I declined them. There followed a melon which my hostess had most appetizingly salted for me with her own hands. I declined again. Then came oatmeal, an indescribable paste on which people might fatten geese in Vienna—then again, perhaps not, since I doubt whether Viennese geese would be willing to eat it. I had already noticed the displeased look of the Alma Mater [Mrs. Hearst] when I refused the melon. Even an Alma Mater is proud of her cooking. Therefore I retched with my head turned away and, thank God, I was not actually sick. That is the unpleasant thing about accepting invitations in America. In hotels one may leave what one cannot eat, but what can one do when faced with a housewife who is proud of the high quality of American cooking in general and her own in particular? Fortunately poultry, compôte and various other things followed with which I could cover the taste of oatmeal.[4]

After managing to somehow survive that gustatory assault (What greater horror could there be than blackberries, melon, and oatmeal?), Boltzmann followed the rest of the dinner party to the music room, where he may have redeemed himself somewhat. There he was invited to play the piano, which he described as "a Steinway from the most expensive price-range."[5] Boltzmann seized the opportunity. He loved to play the piano and, by his own admission, had never before played an instrument "with such beautiful tone."[6] The possibility that his hostess and fellow guests may have also enjoyed his playing he would have considered an incidental benefit. From his perspective, they lived too far from the cultural centers of Europe to have constituted a respectable audience. He was especially dismissive of a music professor from Milwaukee who received this faint praise: "He [the music professor] knew that Beethoven wrote nine symphonies and that the ninth is the last."[7]

If at this point an unflattering picture of our protagonist seems to have emerged, we might restore your sympathies toward him by offering a story in which the joke (by his own telling) was at his own expense. To that end, we turn to his struggles with the English language. Boltzmann was invited to Berkeley to give a series of lectures, which he could have delivered (without any shame) in German. Many of the attendees, in fact, would have been fluent in German. But Boltzmann insisted in attempting to communicate with his audience in English despite the fact that he was no linguist, which he discovered immediately upon his arrival to Berkeley. He described a conversation with a member of his welcoming party (a conversation surprisingly again about food) as if it were a part of some long-lost comedy routine.

This is the form my English conversation took:
I: *When will lunch be served?*
He: *ieeöö.*
I: *I beg you, could you say me, at what hour lunch will be served?*
His splutterings sink a good fifth deeper: *aoouu.*
I realize the error in my plan of attack and cry despairingly: *lönch, lanch, lonch, launch* and so on. I produce vowels which would never be found in Gutenberg's type-cast. Now his face shows a glimmer of understanding: *ah, loanch?* Now the bridge of comprehension has been built:
I: *When? At what hour? When o'clock?*
He: *Half-past one!*
We have understood each other. And now I was supposed to give thirty lectures in this language![8]

Thirty-some lectures later, however, Boltzmann was actually quite pleased with himself. He was especially proud of the way in which he had commanded use of the words *blackboard* and *chalk*, which were difficult for his "protesting tongue" to pronounce.[9] Given the fact that his native language was so generously sprinkled with umlauts, it's not entirely clear why the words *blackboard* and *chalk* would have posed any particular problem, but let us not be sidetracked by an inability (my inability) to recognize which diphthongs were lost in translation. Let us focus instead on Professor Boltzmann, who provided the following self-assessment: "During the first lecture I was somewhat timid, but by the second I was more relaxed and when I finally heard that the students could understand me well, indeed found my presentation lucid and distinct, I soon felt at home."[10]

One might conclude from this self-assessment that Boltz-mann left California with the impression that he had actually made a good impression. But self-assessments are of course not always reliable. Many, if not most, teachers believe themselves to be perfectly well understood, often right up until the moment that they put red pen to paper. It is a delusion of the trade. But Boltzmann may have harbored this delusion to an even greater extent than expected. If he had conducted a more thorough survey of his Berkeley classroom, he would have found (much to his dismay) that his presentations were not universally lauded as lucid and distinct. He would have heard that his language skills were deemed "somewhat deficient, to put it mildly" by at least one member of the audience,[11] and he would have discovered that his tendency to exclaim, "Dass ist the truth!" struck many of his students as an odd (and off-putting) interjection.[12] But inconvenient truths would not have served him well in California, and he may have actually been wiser in old age than we've credited him so far. He may have actually cultivated a willful ignorance later in life that could have served as an effective coping mechanism. Or he may have simply been selectively self-critical and cared not enough about his American audience to have granted them any emotional power over him. He may have reserved that authority for his European colleagues.

His anxiety, in any case, seems to have surfaced in some—but not all—social settings, which would suggest that some mental switch needed to be flipped in order for his illness to prevail. That switch may have been stuck in the on position when conducting business in Europe but not in the United States. One might conclude, therefore, that Boltzmann was something of a snob. One might wish that he had suffered in a more universal and less discriminatory way. But if Boltzmann was a snob, he was a harmless snob who seems to have been more clueless than he was arrogant. And in a great many instances, he clearly suffered from too much, rather than too little, empathy. If, for example, he knew that a student struggled financially, he considered that student's academic success to be his own personal responsibility, and he struggled with the submission of grades more generally because it implied an inevitable need to fail some of his students. In the last year of his life, he failed not a single student on any exam,[13] which his students at the time no doubt appreciated. But we would be amiss here if we did not question his professional judgment on such matters. Boltzmann was a good teacher, but even a good teacher encounters the occasional (or not-so-occasional) bad student. The only way that a teacher can avoid the submission of failing grades is to pass failing students, which in the end is detrimental to all, both to those who've earned their passing grades and to those who have not. We might suspect, therefore, that Boltzmann was driven by good intentions, but he had a professional obligation to manage unpleasant outcomes (and the hurt feelings that they sometimes engendered). And in the last year of his life, he failed to meet that obligation.

In Berkeley, however, he was spared any such concerns. He was invited to California to participate in a summer school, not to give a formal course with the attendant need to submit grades. He needed only to give his thirty lectures, collect his honorarium, and return home, which he managed to do both safely and happily, as he indicated in his review of the trip: "California is beautiful, Mount Shasta magnificent, Yellowstone Park wonderful, but by far the loveliest part of the whole tour is the moment of homecoming."[14] And with that sentiment, we arrive at the last year of his life.

A year after conquering the New World, Boltzmann traveled again, but this time a much shorter distance, to the coastal village of Duino, Italy, where he ended his family vacation by ending his own life. It was not his first attempt at suicide, but it was his most successful. Why suicide? It's impossible to say with any certainty how or why an individual arrives at that decision. But given the amount of collective suffering that exists in the world, it's not entirely self-evident that the will to survive would be so collectively strong or that the decision to live would even be the default position. But it is. We know that every day, most people will decide, consciously or not, not to kill themselves. And there's certainly no shortage of cliffs from which to jump, or poisons to ingest, or firearms to discharge. And there's certainly no shortage of pain or sorrow. So why then do most people, on most days, decide not to kill themselves? Is it fear of the unknown? Or fear of a painful death? Is it a sense of personal responsibility to the loved ones who will remain? Or some unrelenting remnant of hope—that tomorrow might not be as bad as today? Or is it merely a matter of biology—nothing more (or less) than a product of natural selection? All of these factors likely contribute to the persistence of an individual life. But all of these factors combined are still not enough to prevent some individuals from deciding that they've had enough or from simply realizing that they have a say in the matter (that free will allows for its own annihilation).

One might suspect that the decision to die or not to die would also be influenced by a person's religious beliefs, but religious influences no doubt vary because major deities (and their spokespeople) offer conflicting views on the matter. Some traditions would like us to believe that life is always a precious gift, to be preserved at all cost and under any circumstances, and that suffering should be seen as an opportunity to discover one's dignity and give witness to one's faith, while other traditions seem to regard life as an expendable commodity that is happily spent so long as it serves a righteous cause. But for the suicidal person, neither charity nor compassion is to be found in any of these traditions. Why? Because life (especially the end of life) is not always a precious gift; to insist that everyone find dignity in suffering is to hoist a cruel and unusual burden on others; and to kill oneself (and one's enemies) as a statement of any sort is wrong.

So then what are we to do, or think, or say? And what do we make of Boltzmann? We simply do our best. We accept the fact that there are sometimes no right or easy answers, and we try to adopt as sympathetic a view as possible of both ourselves and other people. And for Boltzmann, we wish that he had only been blessed with a more peaceful death, which we imagine for him here.

25 An Alternative Ending

It was a peaceful death. Lying there. He had just eaten dinner—pork and dumplings and one or more glasses of wine. It was a Saturday. After finishing the meal, he retired to the living room with his daughter Elsa and sat down at the family piano. He started to play—something by Schubert—but soon abandoned it, preferring instead to lie down on the couch. He was tired, and with a full and contented stomach, he saw no reason to stay awake, so he settled down on the couch and Elsa took his place at the piano. She asked him what he might like to hear. He suggested the Moonlight Sonata. She smiled at her predictable father and softly played for him its first few arpeggios. He closed his eyes (this for the last time) and allowed his breath to rise and fall with the music—his diaphragm acting as if it were some subtle part of the accompaniment, his thoughts drifting toward microscopic events.

The breath that animated him was a complicated thing, he thought, although he knew it to be nothing more than a gas—something that he had worked his whole life to understand, an ethereal something that forced its way into his lungs with insistent regularity. Had he really understood it? He felt as if he could almost see it, as if he might be some part of it. He thought that maybe he was ready to surrender himself to it, so as the sonata reached its inevitable crescendo, his breath loosened its ties to the music and his body gave way to the inanimate, succumbing to what would become a much higher entropy state. He was a kind man with a generous mind. This was his story and the story of his second law.

Notes

Chapter 2

1. David Lindley, *Boltzmann's Atom: The Great Debate That Launched a Revolution in Physics* (New York: Free Press, 2001), 217.

2. Ibid., 218.

3. Ibid., 231.

Chapter 5

1. Engelbert Broda, *Ludwig Boltzmann: Man-Physicist-Philosopher* (Woodbridge, CT: Ox Bow Press, 1983), 30.

2. David Lindley, *Boltzmann's Atom: The Great Debate That Launched a Revolution in Physics* (New York: Free Press, 2001), 102.

3. Ibid., 101.

4. Ibid.

5. Ibid., 102.

6. Ibid.

7. Ibid., 103.

8. Ibid.

9. Ibid., 104.

10. Ibid., 105.

11. Ibid., 106.

12. Ibid.

Chapter 10

1. David Lindley, *Boltzmann's Atom: The Great Debate That Launched a Revolution in Physics* (New York: Free Press, 2001), 191.

2. Ibid., 185.

Chapter 13

1. Ludwig Boltzmann, "Über die Beziehung zwischen dem zweiten Hauptsatze der mechanischen Wärmetheorie und der Wahrscheinlichkeitsrechnung respektive den Sätzen über das Wärmegleichgewicht," *Wiener Berichte* 76 (1877): 373–435.

Chapter 14

1. Carlo Cercignani, *Ludwig Boltzmann: The Man Who Trusted Atoms* (Oxford: Oxford University Press, 1998), 11.

2. David Lindley, *Boltzmann's Atom: The Great Debate That Launched a Revolution in Physics* (New York: Free Press, 2001), 57.

Chapter 15

1. David Lindley, *Boltzmann's Atom: The Great Debate That Launched a Revolution in Physics* (New York: Free Press, 2001), 198.

2. Engelbert Broda, *Ludwig Boltzmann: Man-Physicist-Philosopher* (Woodbridge, CT: Ox Bow Press, 1983), 12.

3. Ibid.

Chapter 16

1. Engelbert Broda, *Ludwig Boltzmann: Man-Physicist-Philosopher* (Woodbridge, CT: Ox Bow Press, 1983), 11.

2. Ibid.

3. Ibid., 13.

4. Carlo Cercignani, *Ludwig Boltzmann: The Man Who Trusted Atoms* (Oxford: Oxford University Press, 1998), 26.

5. Broda, *Ludwig Boltzmann: Man-Physicist-Philosopher*, 19.

6. Ibid.

7. Ibid., 7.

8. David Lindley, *Boltzmann's Atom: The Great Debate That Launched a Revolution in Physics* (New York: Free Press, 2001), 198.

9. Ibid., 190.

10. Ibid., 184.

11. Ludwig Boltzman, "A German Professor's Journey into Eldorado," *Annals of Nuclear Energy* 4 (1977): 150, translated by Margaret Malt from *Populäre Schriften* (Leipzig: J. A. Barth, 1905).

12. Lindley, *Boltzmann's Atom*, 224.

Chapter 18

1. Stephen Hawking, *A Brief History of Time: From the Big Bang to Black Holes* (New York: Bantam Books, 1988), vii.

Chapter 19

1. Ludwig Boltzmann, "Über die Beziehung zwischen dem zweiten Hauptsatze der mechanischen Wärmetheorie und der Wahrscheinlichkeitsrechnung respektive den Sätzen über das Wärmegleichgewicht," *Wiener Berichte* 76 (1877): 373–435.

2. Max Planck, "Ueber das Gesetz der Energieverteilung im Normal-spectrum," *Annalen der Physik* 309 (1901): 553–563.

3. Max Planck, *Scientific Autobiography and Other Papers*, trans. Frank Gaynor (New York: Philosophical Library, 1949), 41.

4. Malcolm Longair, *Theoretical Concepts in Physics: An Alternative View of Theoretical Reasoning in Physics* (Cambridge: Cambridge University Press, 2003), 339.

5. "Max Planck—Nobel Lecture: The Genesis and Present State of Development of the Quantum Theory," Nobel Media AB, 2014, http://www.nobelprize.org/nobel_prizes/physics/laureates/1918/planck-lecture.html.

6. Planck, *Scientific Autobiography*, 33.

7. Ibid., 32.

8. "Max Planck—Nobel Lecture."

9. P. W. Bridgman, *The Nature of Thermodynamics* (Cambridge, MA: Harvard University Press, 1941), 3.

10. Ibid.

Chapter 20

1. Lewis Campbell and William Garnett, *The Life of James Clerk Maxwell,* 2nd ed. (London: Macmillan, 1884), 97.

2. David Lindley, *Boltzmann's Atom: The Great Debate That Launched a Revolution in Physics* (New York: Free Press, 2001), 183.

3. Ibid., 68.

4. Ibid.

5. Ibid., 66.

6. Engelbert Broda, *Ludwig Boltzmann: Man-Physicist-Philosopher* (Woodbridge, CT: Ox Bow Press, 1983), 28.

7. Carlo Cercignani, *Ludwig Boltzmann: The Man Who Trusted Atoms* (Oxford: Oxford University Press, 1998), 27.

8. Ludwig Boltzmann, *Lectures on Gas Theory* (New York: Dover, 1995), 216, translated by Stephen G. Brush from *Vorlesungen über Gastheorie* (Leipzig: J. A. Barth, 1896, 1898).

9. Lindley, *Boltzmann's Atom*, vii.

10. Ludwig Boltzmann, "Reply to Zermelo's Remarks on the Theory of Heat," in Stephen G. Brush, *The Kinetic Theory of Gases: An Anthology of Classic Papers with Historical Commentary*, edited by Nancy S. Hall (London: Imperial College Press, 2003), 393, translated by Stephen G. Brush from "Entgegnung auf die wärmetheoretischen Betrachtungen des Hrn. E. Zermelo," *Annalen der Physik* 57 (1896): 773–784.

11. Lindley, *Boltzmann's Atom*, 128.

12. Cercignani, *Ludwig Boltzmann: The Man Who Trusted Atoms*, 215.

13. Lindley, *Boltzmann's Atom*, 194.

14. Ibid., 212.

Chapter 22

1. Ludwig Boltzmann, "Reply to Zermelo's Remarks on the Theory of Heat," in Stephen G. Brush, *The Kinetic Theory of Gases: An Anthology of Classic Papers with Historical Commentary*, edited by Nancy S. Hall (London: Imperial College Press, 2003), 397, translated by Stephen G. Brush from "Entgegnung auf die wärmetheoretischen Betrachtungen des Hrn. E. Zermelo," *Annalen der Physik* 57 (1896): 773–784.

2. Carlo Cercignani, *Ludwig Boltzmann: The Man Who Trusted Atoms* (Oxford: Oxford University Press, 1998), 142.

3. Ibid., 141.

4. David Lindley, *Boltzmann's Atom: The Great Debate That Launched a Revolution in Physics* (New York: Free Press, 2001), 82.

5. Ibid., 83.

Chapter 23

1. Rudolf Clausius, "On a Modified Form of the Second Fundamental Theorem in the Mechanical Theory of Heat," *Philosophical
Magazine* 12, no. 77 (1856): 86.

Chapter 24

1. Ludwig Boltzman, "A German Professor's Journey into Eldorado,"
Annals of Nuclear Energy 4 (1977): 147, translated by Margaret Malt
from *Populäre Schriften* (Leipzig: J. A. Barth, 1905).

2. Ibid., 151.

3. Ibid.

4. Ibid., 153.

5. Ibid., 154.

6. Ibid.

7. Ibid.

8. Ibid., 150.

9. Ibid., 151.

10. Ibid., 150.

11. David Lindley, *Boltzmann's Atom: The Great Debate That Launched
a Revolution in Physics* (New York: Free Press, 2001), 205.

12. Ibid., 204.

13. Carlo Cercignani, *Ludwig Boltzmann: The Man Who Trusted Atoms*
(Oxford: Oxford University Press, 1998), 13.

14. Boltzmann, "A German Professor's Journey," 159.

Index

Printed in the United States
by Baker & Taylor Publisher Services